大飞机出版工程

总主编　顾诵芬

英汉航空发动机技术词汇手册

English-Chinese Handbook of Aero Engine Technology

王光秋　编著

上海交通大学出版社
SHANGHAI JIAO TONG UNIVERSITY PRESS

内容提要

本手册主要收集和编辑了航空发动机领域常用的英语技术词汇、专业术语和缩略词共计11 000多条,包括近20年来随着航空发动机新技术发展而衍生出的一些新词汇如blisk、bling等,缩略词如VCE、TBCC和CMC等。所收集的英语技术词汇和缩略词大多数引自英国、美国等国家的航空发动机制造商和科研机构的报告和出版物,涉及发动机的设计、开发、生产、制造和维修等技术领域。这本手册内容简练、词汇丰富、实用性强,适合具有一定英语基础并从事航空产业的工程技术人员以及航空院校的师生使用。

图书在版编目(CIP)数据

英汉航空发动机技术词汇手册/王光秋编著.—上海:上海交通大学出版社,2021.12
大飞机出版工程
ISBN 978-7-313-25987-5

Ⅰ.①英… Ⅱ.①王… Ⅲ.①航空发动机—词汇—英、汉 Ⅳ.①V23-61

中国版本图书馆CIP数据核字(2021)第262918号

英汉航空发动机技术词汇手册
YING-HAN HANGKONG FADONGJI JISHU CIHUI SHOUCE

编 著 者:王光秋
出版发行:上海交通大学出版社　　　　　地　　址:上海市番禺路951号
邮政编码:200030　　　　　　　　　　　电　　话:021-64071208
印　　制:上海万卷印刷股份有限公司　　经　　销:全国新华书店
开　　本:710mm×1000mm　1/16　　　印　　张:23.5
字　　数:459千字
版　　次:2021年12月第1版　　　　　　印　　次:2021年12月第1次印刷
书　　号:ISBN 978-7-313-25987-5
定　　价:188.00元

前　　言

近 20 年来,随着材料科学和先进制造技术的发展,世界航空发动机在产品和技术开发领域都取得了突破性进展。在民用航空领域,涡喷发动机已经基本被淘汰,取而代之的先进大涵道比涡扇发动机已成为在役民用运输类飞机的基本动力装置,具有"革命性技术变革"的齿轮传动发动机也已经成功地进入航线服役。同样,在军用航空领域,用于战斗机的传统涡喷发动机也基本被新一代涡扇发动机所取代;特别是以变循环发动机和自适应变循环发动机为代表的新一代高效航空动力系统已进入验证阶段,预计将被用作美国第五/六代战斗机的推进系统。

随着欧美航空发动机产业和技术的不断发展进步,在航空发动机设计和制造领域也产生和衍生了很多新技术词汇、专业术语和缩略词等,如 GTF(齿轮传动风扇)、VCE(变循环发动机)、TBCC(涡轮冲压组合循环发动机)、blisk(一体化叶片和盘-叶盘)、bling(一体化叶片和环-叶环)、TALON(泰龙低排放燃烧室)和 CMC(陶瓷基复合材料)等。此外,近年来航空发动机制造商在推广新的服务模式中也使用了一些特有词汇,如英国罗罗公司的 TotalCare(包修包维护协议),美国 GE 的 OnPoint(全面维修维护协议)。国内目前现有航空英汉技术词典因编写时代所限,不能反映出近 20 年来航空发动机技术发展而产生的专业技术词汇。编著本英汉航空发动机技术词汇手册主要目的是弥补国内在航空发动机英汉技术词汇领域的空白,与国内已有的航空发动机技术方面的英汉词典形成互补,帮助读者在阅读英文技术资料中查询和理解专业术语和生僻词汇,了解国外相关技术最新发展状况。

本词汇手册主要收集和编写了航空发动机领域的技术词汇、专业术语和缩略语共计 11 000 多条,包括近 20 年来随着航空发动机新技术发展而衍生出的一些新词汇和缩写。本手册所引用的英语词汇和缩略词主要来自英、美航空产业科技报告或科技文章,包括编者本人在英国罗罗航空发动机集团工作 20 年期间收集、整理、编写的大量工作中接触到的专业词汇。编者努力在理解原意和应用技术背景的基础上,给出符合中文习惯的中文翻译。所以,本手册所引用的英文词和中文翻译不仅转自且忠实于原意和技术背景,而且反映出欧美航空发动机技术发展趋势,能够为从事航空发动机产业的科技人员在阅读相关英文技术文献时提供帮助和参考。为了精练内容,减少篇幅,中文翻译原则上仅给出一个对应的汉语译意;此外,书中尽量限制同一个核心词的衍生词组数量,读者可在此基础上举一反三。本手册内容简练、实用性强,可供具有一定英语基础的从事航空发动机研制的工程技术人员以及航空院校的师生使用。

　　本手册获得了西北工业大学精品学术著作培育项目的资助。在本手册的编写过程中,原罗罗公司中国公司秋凡女士帮助输入大量的中文和英文词汇;编者在中国商飞公司北京民用飞机技术研究中心和北京天骄航空产业投资公司任职期间的同事王志峰、蔡烨程、李东杰曾帮助审阅和修改部分稿件;上海交通大学出版社编辑杨小芳审稿严谨,积极提出修改建议;在此一并表示衷心感谢。最后,对编者家庭特别是夫人杨明明给予的全力支持和关照谨表诚挚的感谢!

　　鉴于编者的水平和能力,书中错误和不妥之处在所难免,恳请读者批评指正。

编写说明和凡例

1. 为方便读者便捷查询，所选英语词汇分为"技术词汇"和"缩略语"两大部分。

2. 手册技术词汇中的每个词条包括完整的英文词组全称和中文翻译，除个别习惯用法外，不含缩略语；在缩略语部分中，每个词条均包含三个部分：缩写、英文词组全称和中文翻译。

3. 在手册正文中，一般英文词汇首字母均小写，人名、地名、专有词汇、习惯用法等首字母大写。

4. 缩写词汇构成的缩略语一般用大写字母拼写，通常由全部词汇的首个字母组成，但也有例外，如 PLASMAER，英文全称为 useful plasma for aerodynamic。全称中有缩略语也用大写表示，机构、单位等每个词首字母大写，其余全称中每个字母均为小写。

5. 通常两个英文单词间空一格或加"-"，但专有词汇例外，如罗罗公司的TotalCare、GE 的 OnPoint 等。

6. 词条按英文字母顺序排列，不考虑大小写，排列时也不考虑"/"、"-""()"等符号。

7. 词条前带有数字的，则按数字顺序排列，如 3D blading concept；词条后带有数字的，则按数字顺序排在相应字母之后，如 Inco 718。

8. 括号()用法：表示可以换用，或表示补充说明，或表示省略部分，如 Package B 发动机升级包 B(罗罗)，说明是罗罗公司的专属服务项目；()放在中文翻译之前，其内容表示所属国家或额外说明，如 Panther(以色列)"黑豹"无人机。

目　　录

技 术 词 汇

150 hours endurance test	150 小时持久性试车
1st Eng. Unit	首套发动机
24×7 AOG support	每周 7 天每天 24 小时飞机停航待修支持
2-row (dilution less) porting standard	两排稀释进气孔
3 bearing swivel module	3 轴承旋转单元
3-bearing swivel duct	3 轴承旋转喷管
3D blading concept	三维叶片设计方案
3D coordinate measuring machine	三维坐标测量机
3D resin transfer molding	三维树脂传递模塑成型
3D sensing system	三维扫描测量仪
3D woven resin transfer molding	三维编织树脂传递模塑成型
3-stream burner	三股气流喷嘴
3-stream design	三股气流设计
4 door thrust reverser	四门反推装置
6 seat abreast	每排 6 座(飞机客舱)
6th EU FRP	欧盟第 6 框架项目

a long-awaited follow-on order	期待已久的后续订单
A. V. Roe and Company	(英国)阿芙罗飞机公司
ablative heat shielding material	烧蚀防热材料
aborted take-off	因事故中断起飞
aborted take-off	中断起飞
above mean sea level	高出平均海平面
abradable liner seal	耐磨内衬密封
abradable lining	摩擦衬环
abradable technology	耐磨工艺
abradable thickness	耐磨层厚度
abrasion	擦伤
abrasion blast cleaning	喷砂清理
abrasion cutting	磨削

abrasion failure	磨损失效
abrasion machining process	研磨加工
abrasion resistance	耐磨性
abrasion resistant	抗磨损
abrasion wheel	砂轮
abrasive	磨料
abrasive disc	金刚砂磨盘
abrasive flow machining	磨粒流加工
abrasive hardness	耐磨硬度
abrasive tool	磨具
abrasive waterjet	磨料水射流
absolute altitude	海拔高度
absolute ceiling	绝对升限
absolute pressure	绝对压力
absolute temperature (Kelvin)	绝对温度（K）
absolute traffic capacity	最高交通容量
absorbent filter	吸收式过滤器
accel	加速
accel schedule	加速调节方案
accel time requirement	加速时间要求
accel transient	加速瞬态
accelerated improvement workshop	精益管理（波音）
accelerated mission test	加速任务试车
accelerated take-off	助推起飞
accelerate-stop distance available	中断起飞可用场长度
accelerating the electrification of flight	快速电动飞机项目（罗罗）
accelerating valve	加速活门
acceleration control	加速控制
acceleration control unit	加速控制装置
acceleration distance	加速距离
acceleration time	加速时间
accelerometer	加速度计
accept on deviation	（特例接受有超差的部件）特采
acceptable deferred defect	可延缓待修故障

acceptable means of compliance	适航性验证方法
acceptable quality level	合格质量标准
acceptance (production) test schedule	（产品)验收测试计划
acceptance criteria	验收准则
acceptance flight	验收试飞
acceptance standard	验收标准
acceptance test procedure	验收测试步骤
acceptance test schedule	验收试车计划
access hatch	探视窗口
access ladder	登机梯
access panel	口盖
access port	维护孔
accessibility	可达性
accessory control specification	附件控制规范
accessory drive	附件传动
accessory drive pad	附件传动装置安装座
accessory drive system	附件传动系统
accessory drive torque	附件传动扭矩
accessory gearbox	附件传动箱
accessory pad	附件安装座
accessory placement on gearbox	齿轮箱上的附件布局
accident investigation	事故调查
accident rate per million	每百万次飞行事故率
accident telex	事故通报
accounting period	会计年度
accreditation board	认证委员会
accumulated profit	累计利润
accumulated tolerance	累积公差
acid cleaning	酸清洗
acid dip	酸浸
acid proof alloy	耐酸合金
acieration	渗碳
acoustic fatigue	音波疲劳
acoustic liner	声衬

acoustic lining	降噪内衬层
acrobatic flight	特技飞行
across corner view	对角视图
actions	举措
active aeroelastic wing	主动气动弹性机翼
active clearance control	主动间隙控制
active combustion control	主动燃烧控制
active control technique	主动控制技术
active cooling for hot section	热端部件主动控制冷却
active countermeasures	主动反制
active damper shut off valve	主动阻尼截止阀
active duty	现役
active fan stability augmentation	主动控制风扇增稳
active jet noise control	喷气噪声主动控制
active magnetic bearing	主动电磁轴承
active noise control	噪声主动控制
active performance seeking control	主动寻找并控制最佳性能状态
active surge margin control	主动喘振裕度控制
active tip-clearance control	主动叶尖间隙控制
active vibration control	主动振动控制
active visual stealth material	主动视觉隐身材料
active combustion high-velocity air-fuel spraying	活性燃烧高速燃气喷涂
activity based costing	作业成本法
actual aircraft movement rate	实际飞机起降率
actual assets	实际资产
actual cash value	实际现金价值
actual hours to date	实际小时数
actual time of departure	实际离开时间
actuating sleeve	驱动套筒
actuator element	执行元件
actuators	作动器
adaptive airfoil shape control	自适应翼形控制
adaptive control	自适应控制

adaptive control of engine stall margin	发动机失速裕度自适应控制
adaptive cycle engine	(美国)适应性循环发动机
adaptive engine control	发动机自适应控制
adaptive engine control system	自适应发动机控制系统
adaptive engine technology development	(美国)自适应发动机开发项目
adaptive engine transition program	(美国)自适应发动机转换项目
adaptive maintenance	适应性维修
adaptive rotor blades	自适应旋翼
adaptive versatile engine technology	(美国)自适应通用发动机技术项目
addeneum line	齿冠线
additional test	附加试车
additive drag	附加阻力
adherence to customers	客户第一
adhesive film	胶黏剂薄膜
adhesive wear	粘着磨损
adiabatic	绝热的
adiabatic change	绝热变化
adiabatic change of state	绝热状态变化
adiabatic compression	绝热压缩
adiabatic compression flow	绝热可压缩流
adiabatic efficiency	绝热效率
adiabatic expansion	绝热膨胀
adiabatic flame temperature	绝热火焰温度
adiabatic gradient	绝热梯度
adiabatic lapse rate	绝热温度递减率
adiabatic reversible process	绝热可逆过程
adiabatic system	绝热系统
adjustable bearing	可调轴承
adjustable inlet	可调进气道
adjustable pitch propeller	调距螺旋桨
adjustable shaft bearing	可调轴承
adjuster	调节器
adjusting bolt	调整螺栓
Adour	阿杜尔发动机(赛峰集团,法国)

advance clearance	预办离港手续
advance warning sign	预警标志
advanced aeroengine materials	(英国)先进航空发动机材料项目
advanced affordable turbine engine	(美国)先进经济可承受涡轮发动机计划
advanced airborne vibration monitor	先进的航空振动监测仪
advanced airbreathing propulsion	先进吸气推进器
advanced coating concepts	先进的涂层方法
advanced combat experimental	技术验证机(达索)
advanced combustor technology programme	先进的燃烧室技术开发计划
advanced composite cargo aircraft	先进复合材料货运飞机
advanced control technology for integrated vehicles	一体化飞行器先进控制技术
advanced controls and monitoring	先进的监控
advanced cycles	先进的循环
advanced demonstrator program	技术验证项目
advanced design methodology	先进的设计方法
advanced design principles and processes	先进设计原理与流程
advanced design tools	先进的设计工具
advanced drawing issue	先期出图
advanced ducted propulsion	先进涵道推进
advanced ducted propulsor	先进涵道推进器
advanced engineering memorandum	先进工程备忘录
advanced european jet pilot training	先进欧洲喷气机驾驶员培训
advanced gas reactor	先进燃气发生器
advanced gas turbine	先进燃气涡轮
advanced general aviation transportation experiments	先进通航运输试验
advanced high pressure turbine blade tip cooling	先进的高压涡轮叶片端冷却
advanced high temperature engine materials technology program	先进发动机高温材料技术项目
advanced ignition	提前点火
advanced integrated manufacturing system	先进的组合式制造系统

advanced jet trainer	先进喷气式教练机
advanced light combat aircraft	(捷克)先进轻型攻击机
advanced low emissions combustion systems	先进低排放贫油燃烧系统项目（罗罗）
advanced medium STOL transport	(美国)先进中型垂直起降运输机
advanced multi-role tanker transport	先进多功能空中加油机
advanced on condition monitoring	先进的状态监视
advanced product design	先进产品设计
advanced product quality planning.	先进产品质量计划
advanced prop-fan engine technology	先进桨扇发动机技术
advanced quality planning	先进质量计划
advanced replacement	先期更换
advanced spectrum analysis	先进的频谱分析
advanced stall	深度失速
advanced subsonic transport	先进亚声速运输机
advanced supersonic propulsion and Integration Research	先进超声速推进与集成研究
advanced supersonic transport	先进超声速运输机
advanced tactical fighter	先进战术战斗机
advanced technology fan integration	一体化先进技术风扇
advanced transport technology	先进运输技术
advanced turbine engine company	先进涡轮发动机公司(霍尼韦尔与普惠合资)
advanced turbine engine gas generator	先进涡轮发动机燃气发生器
advanced turbine rotor durability	先进涡轮转子耐久性
advanced turbine technology applications	先进涡轮技术应用
advanced turbo-prop	先进涡桨发动机
advanced turboprop project	先进涡桨项目
advancing blade	前行桨叶
advancing blade concept	(直升机)前行桨叶方案
advancing blade stall	前行桨叶失速
advisory airspace	咨询空域
advisory circular	咨询通告

Advisory Council for Aeronautic Research in Europe　欧洲航空研究咨询委员会

Advisory Group for Aerospace Research and Development　航空研发咨询委员会

advisory service　咨询服务

AerCap　(总部位于荷兰的一家全球性综合航空公司)爱尔开普

aerial common sensor　空中通用传感器(美国侦察机)

aerial propeller　空气螺旋桨

aerial propeller vessel　空气螺旋桨推进船

AERMACCHI　(意大利)马基飞机制造公司

aero design feature　气动设计特性

aero engine design principles　航空发动机设计原理

aero engine technology　航空发动机技术

aeroacoustics　气动声学

aerobatics　特技飞行

aerobatics maneuver　特技飞行动作

aero-derivative　航空衍生的

aero-derivative gas turbine　航改型燃气涡轮机

aero-derivative power turbine　航改型动力涡轮

aerodrome　小型飞机场

aerodrome control service　机场管制服务

aerodrome control tower　机场管制塔台

aerodrome obstruction chart　机场障碍图

aerodromometer　流速表

aerodynamic "A" frame　有气动特性的"A"形框架

aerodynamic buffeting impulses　空气动力抖动冲击

aerodynamic centre　气动力中心

aerodynamic damping　气动阻尼

aerodynamic efficient　气动效率

aerodynamic interference　气动干扰

aerodynamic load of engine　发动机气动载荷

aerodynamic tailoring　气动裁剪

aerodynamically induced vibration　空气动力诱导的振动

aeroelastic effect	气动弹性效应
aero-engine	航空发动机
aerofoil super polishing	叶身超抛光
aerogram	高空气象图
AEROJET	(美国)航空喷气公司
aeromanager	(罗罗北美数字化管理系统)航空管理器
aeronautical engineer	航空工程师
aeronautical engineering	航空工程
aeronautical radio incorporated	一体化航空无线电
aeronautical systems division	航空系统部
aeronautics	航空学
Aeronautics Research Mission Directorate	航空科研规划局
aero-propeller	航空螺旋桨
aeroservoelastic charateristic	气动弹性特征
aerosol	喷雾剂
aerosonator	脉冲式喷气发动机
Aerospace and Defence Industries Association of Europe	欧洲航空航天与防务工业协会
Aerospace Foundries	航空产品铸造厂
Aerospace Industrial Development Corp.	(台湾)航空航天工业发展集团
Aerospace Industries Association of America	美国航空航天工业协会
Aerospace Industry Reports	航空航天工业报告
Aerospace Material Specifications	航空航天材料规范
aerospace propulsion technology demonstrator	(中国)航空推进技术验证项目
aerospace quality systems	航空航天质量体系标准
aerospace recommended practices	航空航天操作规程建议
aerospace standard for quality management	航空航天质量管理
aerospace standard(s)	航空航天标准
Aerospace Technology Institute	(英国)航空技术研究所
Aerospatiale	(法国)法宇航公司
aerothermodynamics	气动热力学
aerotoxic syndrome	航空中毒综合征
aero-turbine	航空涡轮机

affordable near term low emissions	（欧洲）可承受的近期低排放发动机项目
"A" frame mount attachment	"A"形安装固定架
aft fan	后风扇
aft movement	向后移动
aft pylon fairing	后挂架整流罩
aft reverser	后反推力装置
aft thrust reverser controls	后反推装置控制
after cooler	后冷却器
after propellor aircraft	后置桨扇飞机
after take-off	起飞后
afterbody drag	尾部阻力
afterburner	加力燃烧室
afterburner blowout	燃烧室熄火
afterburner casing	加力燃烧室筒体
afterburner combustion efficiency	加力燃烧效率
afterburner diffuser	加力燃烧室扩压器
afterburner fuel distribution test	加力燃油浓度分布试验
afterburner fuel manifold	加力燃油总管组件
afterburner fuel supply in multi-zone and staged pressure manner	加力室多区分级压力法供油
afterburner igniter	加力燃烧室点火器
afterburner ignition	加力燃烧室点火
afterburner liner	加力燃烧室耐热层
afterburner model test	加力燃烧室模型试验
afterburner operating condition	加力燃烧室工作状态
afterburner sector test	加力燃烧室扇形段试验
afterburner stability	加力稳定性
afterburner thrust gain	加力推力增益
afterburner total pressure recovery coefficient	加力燃烧室总压恢复系数
afterburning temperature	加力温度
aftermarket purchase services	售后采购服务
after-market revenues	售后市场收入
Agencia Nacional de Aviacao Civil	巴西国家民航局

agile manufacturing	敏捷制造
agility	(飞机)机动性
aids to navigation	辅航设备
Air Accidents Investigation Branch	(英国)航空事故调查局
air ambient temperature	外部空气温度
air annulus	气流环形通道
air blast	鼓风
air blast atomizer	空气雾化喷嘴
air blast quenching	强风淬火
air bleed system	放气系统
air bleed valve	放气活门
air brake	空气制动器
air breakaway	空气分离
air breathing engine	吸气式发动机
air care	空中救援
Air Cargo Management Group	世界航空货运管理集团
air cargo sector	航空货运业
air cavity system	(新型)气垫船系统
air composition	空气成分
air control valve	气控阀
air cooled oil cooler	气冷滑油冷却器
air cooler	空气冷却器
air corridor	空中走廊
air cushion pressure	气垫压力
air data computer	大气数据计算机
air display team	飞行表演队
air duct	空气导管
air en route control	航路管制
air fares	航空票价
air film seal	气膜密封
air flow control	空气流量控制
Air Force 1st Research Institute	空军第一研究所
Air Force of Chinese People's Liberation Army	中国人民解放军空军
Air Force Materiel Command	美国空军装备总部

air freight	航空货运
air fuel ratio	空气与燃油比
air heater	空气加热器
air impingement starting system	空气冲击起动系统
air induction system	进气系统
air injection pressure	空气喷射压力
air inlet control system	进气道控制系统
air inlet control unit	进气道控制单元
air inlet slots	进气槽
air intake	进气道
air intake test	进气道试验
air inter cooler	中间空气冷却器
air jet distortion generator	空气射流畸变模拟器
air leakage	漏气
air logistic support	空中后勤支持
air logistics	航空物流
air mass flow	空气量流
air preheater	空气预热器
air pressurization	空气增压
air propeller	空气螺旋桨
air property	空气性质
air recirculation	空气再循环
air refuelling	空中加油
air relief unit	放气机构
air resistance	空气阻力
air riding carbon seals	气动碳环封严
air separation module	空气分离模块
air speed	相对空气的飞行速度
air spray	喷雾
air starter	空气启动机
air starting	以空气驱动启动
air starting cam	空气起动凸轮
air straightener	空气矫直器
air strainer	空气滤网

air strike	空袭
air superiority	空中优势
air supremacy	制空权
air surveillance and reconnaissance	机载监视与侦察
air system	空气系统
air tapping	气嘴
air tapping system	引气系统
air terminal	航站候机楼
air test	飞行试验
air traffic advisory service	空中交通咨询服务
air traffic beacon	空中交通信标
air traffic categories	空中交通类型
air traffic control	空中交通管制
air traffic control centre	航空交通控制中心
Air Traffic Control Committee	空中交通管制委员会
air traffic development	空中交通的发展
air traffic flow and capacity management	空中交通流量与容量管理
air traffic flow management	空中交通流量管理
air traffic forecast	航空交通预测
air traffic management	空中交通管理
air traffic organization	空中交通组织
Air Transport Association	空运协会
Air Transport Licensing Authority	空运牌照局
air turbine starter	空气涡轮启动器
air turbine starter	空气涡轮起动机
air/oil cooler	以空气为冷媒的滑油冷却器
air/oil heat exchanger	油气热交换器
airblast atomizer	空气雾化喷嘴
airblast fuel nozzle	燃料喷嘴
airblast nozzle	喷气嘴
airborne collision avoidance system	机载防撞系统
Airborne Command and Control Centre	机载战场指挥控制中心
airborne early warning	空中预警
airborne early warning & control system	空中预警与控制系统

airborne equipment	机载设备
airborne system	机载系统
airborne vibration monitoring system	机载振动监视系统
airborne warning and control system	空中预警与控制系统
airbrake	减速板
airbreathing propulsion system	吸气式推进系统
airbreathing pulsed detonation engine	吸气式脉动爆燃发动机
Airbus	空客公司
airbus corporate jetliner	空客公务喷气机
airbus flight operational commonality family	空客飞行运营通用家族
Airbus SE	空客集团(德/法/西)
Aircelle	(赛峰生产短舱的子公司)埃塞公司
air-cooled blade	气冷工作叶片
air-cooled compressor	气冷压气机
air-cooled engine	气冷发动机
air-cooled generator turbine	气冷涡轮发电机
air-cooled heat exchanger	气冷式热交换器
aircraft due by	至……年飞机到役
aircraft angle of attack	飞机迎角
aircraft approach procedure	飞机进场程序
aircraft base maintenance	飞机基地维修
aircraft certification	飞机取证
aircraft characteristics	飞机特性
aircraft communications addressing and reporting system	机载通讯寻址和报告系统
aircraft component release certificate	航空器部件放行证书
aircraft deferred defects	飞机延误故障
aircraft drag coefficient	飞机阻力系数
aircraft energy efficiency	飞机能耗效率
aircraft engine control	飞机发动机控制
aircraft engine repair overhaul	飞机发动机检修
aircraft engine research laboratory	飞机发动机研究实验室
aircraft engine test	飞机发动机测试
aircraft fleet projection	飞机机队预测

aircraft hangar	飞机机库
aircraft identification	飞机识别
aircraft integrated data system	飞机综合数据系统
aircraft integration	飞机一体化
aircraft integration definition plan	飞机一体化计划
aircraft lift coefficient	飞机升力系数
aircraft maintenance management system	飞机维护管理系统
aircraft maintenance manual	飞机维护手册
aircraft maintenance support services	飞机维修支持系统
aircraft mounted accessory drive	机上附件传动装置
aircraft on ground	飞机停航待修
aircraft overdue	飞机超期服役
aircraft performance	飞机性能
aircraft power feeder cable	飞机电源馈线
aircraft power on	飞机通电测试
aircraft prepared for service	待命飞机(不含燃料和载荷)
aircraft propulsion system	飞机推进系统
aircraft propulsion system design optimization	飞机推进系统优化设计
aircraft survivability equipment	飞机生存性设备
aircraft test procedure	机上实验程序
aircraft transient requirement	飞机瞬态性能要求
aircraft utilization	飞机使用率
aircraft vector propulsion technology	飞行器矢量推进技术
aircraft/engine integration design	飞机-发动机一体化设计
airflow	空气流量
airflow control	空气流量控制
airflow measurement	空气流量测量
airflow meter	空气流量计
airflow system	气流系统
airfoil angle variation	叶形角变化
airfoil angle's tolerance	叶形角公差
airfoil categories	叶型分类
airfoil contour change	翼型外形变化
airfoil cooling methods	叶片冷却方法

airfoil count	叶片数
airfoil lean and bow	翼型倾斜和弧形变化
airfoil profiles	叶型
Airfoil Services Kuala Lumpur	吉隆坡发动机叶片服务公司
airframe and powerplant	机身与动力装置
airframe related components	机体结构部件
air-fuel mixture	燃料空气混合
air-fuel ratio	燃料空气混合比例
air-launched cruise missile	空中发射巡航导弹
airless blast cleaning	真空喷砂清理
airline master scheduling team	航空公司交付安排小组
airline on-time performance	航空公司准时性业绩
airliners	班机
airline's carbon footprint	航空公司的碳排放
airplane cruise thrust	飞机巡航推力
airplane health management	机队实时健康管理
airplane performance monitoring	飞机性能监测
airport compatibility	机场的兼容性
airport traffic management	机场交通管理
airspace	空域
airspace control order	空中管制命令
airspace over the territory	领土上空
airspace structure	空域结构
airspeed indicator	空速表
airspray	喷雾
airspray nozzle	空气雾化喷嘴
air-starter assisted relight	空气启动器辅助再点火
air-surveillance jet	空中预警机
air tanker	(英国)空中加油机
air-to-air refuelling	空中加油
airworthiness approval certificate	适航审批证书
airworthiness approval tag	适航批准挂签
airworthiness certification	适航证
airworthiness characteristics	适航特性

airworthiness circular	适航通告
airworthiness data	适航性资料
airworthiness directions	适航说明
airworthiness directive	适航指令
airworthiness flight test	适航性飞行试验
airworthiness flight test schedule	适航性飞行试验进度表
airworthiness hazard	适航性灾难
airworthiness limitation	适航性限制
airworthiness limitation item	适航限制项目
airworthiness manual	适航手册
airworthiness notice	适航管理通告
airworthiness procedures	适航管理程序
airworthiness regulation for commercial engines	民用发动机适航性条例
airworthiness standard on aircraft noise	噪声适航标准
airworthiness supervision	适航性检查
alarm signal	警报信号
alarm thermometer	温度计报警
alarm valve	警报阀
Albatro	信天翁-教练机(沃多乔迪航空公司, 捷克)
Al-diffusions bond layer	铝扩散连接层
Alenia Aermacchi	(意大利)马基飞机公司
Alenia Aeronautica	(意大利)阿莱尼亚航空工业公司
alert service bulletin	报警服务公告
align metrics	优化方法
all composite aircraft	全复合材料飞机
all composite airframe	全复合材料机身
all composite fan containment case	全复合材料风扇包容机匣
all electric aircraft	全电化飞机
all electric engine	全电化发动机
all operators telex	全体用户通报
All Russia Institute of Aviation Materials	(俄罗斯)俄罗斯航空材料研究院
all up weight	飞机总重
all-axis nozzles vectoring thrust	(俄米格公司的)全向推力矢量

allegedly violating	违规
Allied Aerospace	(美国)联合宇航公司
allied concurrent engineering	联合并行工程
allied hazard	同类危险
allied industry	配套工业
Allied Signal	(美国)联信公司
Allison	(美国)艾利逊公司
Allison Advanced Development Company	(美国)艾利逊先进技术开发公司
Allison Engine Company	(美国)艾利逊发动机公司
Allison Engine Company（now Rolls-Royce Corporation）	艾力逊发动机公司(现为罗罗北美公司)
all-metal adhesive bonding structure	全金属胶接结构
all-new carbon fiber	全新碳纤维
allowable pressure	许用压力
Alston	阿尔斯通公司
alt	高度
alternate fuels	替代燃料
alternate hight-frequency material	交变高频材料
alternating current	交流电
alternating stress	交变应力
alternative engine	代用发动机
alternative fuel effects on contrails and cruise emissions	替代燃料对航迹和巡航排放的影响
alternative load path	辅助传力路线
alternator	交流发电机
alternator frequency	交流发电机频率
alternator voltage	交流发电机电压
altitude capability	升限
altitude characteristics	高度特性
altitude component test	高空部件试验
altitude flight	高空飞行
altitude relight	高空点火
altitude sensing unit	高度传感装置
altitude simulating test	高空模拟实验

altitude test	高空试验
altitude test chamber	高空试验台
altitude test facility	高空台
altitude wind tunnel	高空风洞
altitude wind-milling test	高空风车试验
aluminium alloy rear drum	铝合金后轮鼓
aluminium annulus filler	铝材料环型整流片
aluminium casting	铝铸件
aluminium extrusions & sections	铝型材及其制品
aluminium honeycomb core	铝蜂窝芯
aluminium isogrid plate	铝格板
aluminium-magnesium alloy	铝-镁合金
aluminium oxide	氧化铝
aluminium ring	铝环
aluminizing	渗铝
aluminum alloy	铝合金
aluminum-lithium alloy	铝-锂合金
ambient air	外界空气
ambient pressure	大气压力
ambient temperature	环境温度
American Aerospace Quality Group	美国航空航天质量协调组织
American Association for Laboratory	美国实验室协会
American Association for Laboratory Accreditation	美国实验室鉴定协会
American customer satisfaction index	美国客户满意指数
American Institute of Aeronautics and Astronautics	美国航空宇航研究所
American Military Standard	美国军队标准
American National Standards Institute	美国国家标准协会
American Society for Engineering Education	美国工程教育协会
American Society for Testing and Materials	美国材料和试验协会
American Society of Mechanical Engineers	美国机械工程师协会
American Society of Non-Destructive Testing	美国无损检测协会
American Transport Association	美国运输协会

American Welding Society	美国焊接协会
amphibious fire-fighting aircraft	水陆两用灭火飞机
amplifier performance	放大器性能
analog-to-digital transducer	模拟数字转换器
analytical spot size	分辨率
anchor bolt	固定螺栓
anchor plate	固定板
aneroid barometer	膜盒气压表
angle gear box	角齿轮箱
angle of attack	迎角
angle of attack indicator	迎角指示器
angle of attack sensor	迎角传感器
angle of incidence	入射角
angle of inclination	倾斜角
angle of skew back	（螺旋桨）后倾角
angular acceleration	角加速度
angularity coefficient	角度系数
anisotropic material	各向异性材料
annealing furnace	退火炉
annex 16 chapter 3	附录 16 第 3 章
annual balance sheet	年终决算表
annual utilization	年利用率
annular cascade test	环形叶栅试验
annular casing	环型机匣
annular combustion liner cooling ring	环型燃烧室衬套冷却环
annular combustor	环行燃烧室
annular diffuser cascade	环形扩压叶栅
annulus filler	环向填充材料
anti-backlash device	防松装置
anti-backlash gear	无齿隙齿轮
anticipation philosophy	预期方案
anti-competitive state subsidies	违反竞争性的国家补贴
anti-corrosion	防锈
anti-dumping duty	反倾销税

anti-friction paints	增滑漆
anti-ice	防冰
anti-ice duct	防冰管
anti-ice exhaust grid	防冰排气格栅
anti-ice switch	除冰开关
anti-ice valve	防冰活门
anti-icing	防结冰
anti-icing system of engine	发动机防冰系统
anti-infrared interference technology	红外抗干扰技术
Anti-Monopoly Law	反垄断法
anti-rotary dog	制动件
anti-shock mount	防震螺栓
anti-siphon device	防虹吸装置
anti-squeak	消声器
anti-stick	防黏结
antitrust investigation	反垄断调查
Antonov Company	(乌克兰)安东诺夫飞机公司
AOG warehouse	紧急航材仓库
applicable aircraft model	适用机型
application for variance/deviation	变更申请
application of lifecycle management	应用程序生命周期管理
applications specific integrated circuit	特殊集成电路应用
appraisal plan	评估计划
approach	进场,进近
approach speed	进场着陆速度
approach time	进场时间
appropriate radius	恰当半径
approval of the design change	设计更改的批准
approve & maintain	批准与维护
approved flight	批准航班
approved maintenance organisation	经授权的维修机构
approved standard	经批准的标准
approved test certificate	经批准的试验证明书
approved to proceed	项目启动

auxiliary power unit maintenance permit	辅助动力装置维修许可证
aramid-aluminium fiber-metal laminate	聚芳族酰胺纤维铝合金层板
arc furnace	电弧炉
arc heating generator	电弧加热发电机
arc welding	电弧焊
area weighting rakes	面积加权测量耙
argon oxygen decarburization	氩氧脱碳
armature assembly	电枢组件
armco	(低碳软铁)阿姆科
armed helicopter	武装直升机
armed reconnaissance helicopter	武装侦察直升机
arms embargo	武器禁运
Arnold Engineering Development Centre	(美国)阿诺德工程开发中心
around-the-clock accessibility	24 小时的客户支持
Arrius	阿赫尤斯发动机(透博梅卡公司)
arrival metering program	进场计量程序
artificial neutral net	人工神经网络
asbesto	石棉
Asia & SW Pacific	亚洲和西南太平洋
Asia-Pacific Aerospace Quality Group	亚太航空航天质量协调组织
aspect ratio	展弦比
aspheric surface	非球面
assembling bolt	装配螺栓
assembly and test engineering	组装和试验工程
assembly area	组装区
assembly center	总装中心
assembly chart	装配表
assembly fitter	装配工
assessing authorized live	授权现场评估
assessment of engine stability	发动机稳定性评估
asset appraisal	资产评估
assets	资产
assistance start	辅助启动
assisted short take-off and vertical landing	利用助推器短距起飞和垂直着陆

Association of European Airlines	欧洲航空公司协会
Atar	阿塔尔发动机(罗罗公司,英国)
ATENA Engineering GmbH	(德国)阿逊纳工程公司
atmosphere plasma spraying	常压等离子喷涂
atmosphere refraction	大气折射
atmospheric absorption	大气吸收
atmospheric line	大气压力线
atmospheric pressure	大气压力
atom absorbing spectrum meter	原子吸收光谱仪
Atomic Energy Commission	原子能委员会
atomized fuel	雾化燃油
attack helicopter	攻击直升机
attained turn rate	瞬时转弯速率
attempted pump	连体泵
audio electronic control unit	音频电子控制单元
audit certificate	审计证书
audit of financial statement	财务报表审计
auditing	审核
Auger electron spectroscopy	俄歇电子能谱
augmentation ratio	加力比
augmented rating	加力推力设定
augmented reality	增强现实技术
augmented vectored thrust	加力矢量推力
augmentor	加力装置
Augusta Westland	(意大利)阿古斯塔-韦斯特兰公司
Aurora Flight Sciences Corp.	(美国)极光飞机科学公司
austenitic steel	奥氏体钢
Australian Quarantine and Inspection Service	澳大利亚检验检疫局
Australian Transport Safety Bureau	澳大利亚运输安全局
authorized military overhaul facilities	经授权的军工大修设施
authorized release certificate	授权颁发适航证
authorized to offer	获准销售
autoflight system	自动飞行系统
automated commercial system	自动商务系统

automated component mode synthesis	自动部件模态综合法
automated guided vehicle	无人搬运车
automated net control device	网络自动控制器
automated welding system	自动焊接系统
automatic accelerator	自动加速器
automatic alarm receiver	自动警报接收器
automatic detector for harness	线束自动检测仪
automatic diagnosis of engine fault	发动机故障自动诊断
automatic direction finder	自动定向仪
automatic landing system	自动着陆系统
automatic pilot	自动驾驶仪
automatic pitch trim	自动俯仰配平
automatic test equipment	自动测试设备
automatic throttle control	自动油门控制
automation system operation	自动化系统运行
automotive application	机动车辆应用
autonomous aerial cargo utility system	(美国)自主无人空中货运服务系统
autonomous and semi-autonomous vehicle	全自动和半自动车辆
autopilot	自动驾驶仪
autopilot engaged	使用了自动驾驶仪
autothrottle	自动油门
autothrottle system	自动油门系统
auxiliaries	附件
auxiliary fuel tank	副油箱
auxiliary gearbox	辅助齿轮箱
auxiliary inlet door	进气道辅助进气门
auxiliary oil pressure sensor	辅助油压传感器
auxiliary power system	辅助动力系统
auxiliary power unit	辅助动力装置
auxiliary power unit bleed valve	辅助动力装置放气活门
auxiliary power unit control switch	辅助动力装置控制开关
auxiliary power unit generator	辅助动力装置发电机
auxiliary power unit load compressor	辅助动力装置负载压缩器
auxiliary pump	辅助泵

auxiliary turbine-generator set	辅助涡燃发电机组
availability rate	可用率
available on demand	按需供货
available seat kilometers	可用座位公里数
available seats on aircraft per nautical mile	机上座位数/海里
availability	有效性
Avenger	复仇者-攻击机(通用动力/麦道公司,美国)
average effective pressure	平均有效压力
average engine	平均发动机性能
average indicated pressure	平均指示压力
average net debt	平均净债
Aviadvigatel OJSC	(俄罗斯)彼尔姆发动机公司
aviation advisory board	航空咨询委员会
Aviation Capital Group	航空金融集团
Aviation Data Communication Corporation	(中国)民航数据通信有限责任公司
aviation fuel tank farm	机场航空煤油库
aviation gasoline	航空汽油
aviation gasoline independence	不依赖航空汽油
aviation hydraulic oil	航空液压油
aviation kerosene	航空煤油
Aviation Register of the Interstate Aviation Committee of the Commonwealth of Independent States	独联体国家航空委员会航空注册局
aviation restructure initiative	航空业重组建议
Aviation Safety Network	(美国)航空安全网
aviation sub-contracting manufacture	航空转包生产
Aviation Suppliers Association	航空供应商协会
aviation support service	航空支援服务
aviation technology, integration, and operations	航空技术,集成与运营
aviation turbine fuel	航空燃气涡轮发动机燃料
aviation unit maintenance	航空设备维护
AVIC Information Technology Co., Ltd.	金航数码科技公司

Avio Aero	艾维欧航空技术集团公司（GE，意大利）
avionics full duplex switched ethernet	航空电子全双工交换式以太网
avionics testing equipment	机载设备测试装置
Avon	埃汶发动机（罗罗公司，英国）
axial deflection	轴向位移
axial dovetail	轴向燕尾槽
axial extended root	轴向延伸叶片根部
axial flow	轴向气流
axial flow turbine	轴流涡轮
axial front stage	轴向式前级
axial increasing pitch	轴向递增螺距（螺旋桨专用）
axial movement	轴向位移
axial root stator	轴向固定导向器
axial-flow compressor	轴流压气机
axial-radial compressors	轴向—径向组合式压气机
axi-centrifugal compressor	离心式压气机

back pressure turbine	反压涡轮
back to back test	直接对比试验
backsweep	后掠
back-up hydro-mechanical system	液压机械备份系统
backward sweep	后掠
baffle	导流片
baffle plate	折流板
Bahrain Civil Aviation Affairs	巴林民航局
balance sheet	资产负债表
balance sheet ratio	资产负债比率
balanced business portfolio	均衡业务
balanced delivery score	平衡交货计分
balanced derate	均衡递减

balanced quality score	平衡质量记分
balanced rotational speed	均衡转速
balanced score card	平衡记分卡
ball indentation hardness	球压痕硬度
ball thrust bearing	滚珠止推轴承
ballast	压载
ballast tank	压载水柜
bandwidth	频带宽
bank angle	（飞机的）倾角
bar	巴（气压单位）
bar codes	条形码
bare engine	（未装短舱的发动机）裸机
bare engine weight	裸机重量
barometer	气压表
barometric altimeter	气压高度表
barometric altitude	气压高度
barometric flow control unit	气压式流量控制器
barring	盘车
base load	基本载荷
base maintenance	基地维护
base metal	金属母材
base plate	底板
baseline	基准
basic design data	基础设计数据
basic design document	基础设计文件
basic engine	发动机基本型
basic operating weight	基本重量
basis of accounting	会计基础
basket purchase	整套采购
batch number	批号
batch size	批量
battery-power aircraft	电池动力飞机（波音公司，美国）
battlefield light utility helicopter	战场轻型通用直升机
battlefield reconnaissance helicopter	战场侦察直升机

bean power density	电子束功率密度
Beanstandungsmeldung	疵品报告单
bearing	轴承
bearing bush	轴承衬套
bearing cap	轴承盖
bearing chamber	轴承腔
bearing damping	轴承阻尼
bearing failure diagnosis and analysis	轴承故障诊断与分析
bearing force	轴承力
bearing metal	轴承金属
bearing plate	支承板
bearing ring	轴承油环
bearing scavenge oil	轴承回油
bearing span	支承跨距
bearing stand	轴承台
bearing temperature	轴承温度
Beechjet	比奇喷气机-公务机（比奇飞机公司，美国）
Beijing Aero Lever Precision Limited	北京力威尔航空精密机械公司
Bell	（美国）贝尔公司
bellcrank	摇臂
bellmouth	进气唇口
belly landing	以飞机腹部着陆
benchmark performance	标杆业绩
benchmarking	基准
"bench mark" test	"台架"试车
benefit of change	改进的优点
Beriev Aircraft Company	（俄罗斯）别里耶夫飞机公司
Berkut	金雕-战斗机（苏霍伊设计局，俄罗斯）
Bernoulli equation	伯努力方程
best cruise altitude and Mach number	最佳巡航高度和马赫数
best performance retention	最佳性能保持
best-selling engine	热销发动机
beyond economic repair	不值得的维修

beyond visual range	超视距
beyond-visual-range missile	超视距导弹
biaxial stress	双轴应力
bidirectional lock	双向锁
biennial airshow	每两年一次的航展
bifurcation	交叉
bifurcation nose fairing	交叉式整流锥
bigrains	双晶
Bilateral Airworthiness and Safety Agreement	双边适航和安全协议
bill of material	材料单
billet	小钢坯
billet standard quality	标准质量锻坯
bimetallic strip	双金属片
binary alloy	二元合金
biodiesel from algae	从海藻中提取的生物柴油
biofuel	生物燃料
bird ingestion test	吸鸟试验
bird strike	鸟撞
bird strike test on fan rig	风扇部件鸟撞试验
bird surge line	吸鸟喘振边界
Blackbird	黑鸟-侦察机(洛克希德-马丁公司，美国)
blade axial retention	叶片轴向固定
blade coning angle	桨叶锥角
blade cooling	叶片冷却
blade failure	叶片失效
blade fatigue test	叶片疲劳试验
blade fixing arrangement	压气机叶片固定方式
blade flir-tree grinding	叶片枞树型根部磨削加工
blade flutter	叶片颤振
blade harmonic frequency	叶片谐频
blade leading edge	叶片前缘
blade loading limit	叶片载荷极限
blade natural frequency	叶片固有频率

blade natural frequency under rotation	叶片动频
blade outer air seal	叶片外部空气封严
blade passing frequency	叶片通过频率
blade peening	叶片喷丸
blade pitch	叶距
blade pitch angle	桨叶角
blade pitch control unit	桨距控制装置
blade platform	叶片减振凸台
blade process casting	铸造叶片流程
blade process forging	锻造叶片流程
blade process machining	叶片机加工流程
blade profiling	叶片造型
blade retainer	叶片护圈
blade row loss	叶道损失
blade scatter	叶片的散布
blade separation device	叶片分离装置
blade shaping	叶片造型
blade stall	叶片失速
blade steady loading noise	叶片定常负荷噪声
blade strength calculation	叶片强度计算
blade tip hardface coating	叶尖表面硬化涂层
blade tip speed	叶尖速度
blade tip timing	叶尖定时测量
blade tip timing frequency	叶尖定时频率
blade twist angle	叶片扭角
blade unsteady loading noise	叶片非定常负荷噪声
blade vibration	叶片振动
blade with vortex generators	带旋涡发生器叶片
bladed disc	整体叶盘
bladed ring	整体叶环
bladed rotor integrity	一体化叶片和转子
blade-disc coupling vibration	叶盘耦合振动
blast heater	鼓风加热器
blasting grit	喷砂

bleed	放气
bleed air precooler	放气预冷器
bleed air temperature control sensor	放气温度控制传感器
bleed flow	引气量
bleed orifice excitations	放气孔激振
bleed port actuation	放气口传动
bleed systems	放气系统
bleed valve	放气阀
bleed valve control unit	放气阀控制器
bleeding-edge technologies	尚未验证的新技术
bleedless	无引气
bleedless engine	无客舱引气的发动机
blended wing body	翼身融合飞机
blended wing body configuration	翼身融合式布局(飞机)
blended winglet	融合式小翼
blending of radii	圆滑过渡
bling (＝bladed ring)	整体叶片环
bling rotor	无盘转子
blisk (＝bladed disk)	整体叶盘
blisk compressor	整体叶盘压气机
blisk configuration	叶盘构型
blistering	起气泡
block fuel	航程总油耗(轮挡油耗)
block speed	(无风条件)航线平均速度
block test endurance	持久性台架试验
block time	轮档时间(总航时)
blockage of the signal line	堵塞信号线
blockage ratio	堵塞比
blocker door	折流门
blocker doorless thrust reverser	无折流门反推装置
blow-off coefficient	吹除系数
blowout	熄火
blue-print	蓝图
BMW Rolls-Royce GmbH	(德国)宝马-罗罗公司

BODYCOE	（热处理专业公司）鲍迪克
body-fitted grid	贴体网络
Boeing 787 dreamliner	波音 787 梦幻客机
Boeing Company	（美国）波音公司
Boeing propulsion systems division	波音公司推进系统部
boiler	锅炉
boil-off gas heater	蒸发液化气加热器
bolt clippers	螺栓钳
bolt head	螺栓头
bolt joint	螺栓连接
bolt point	螺栓尖
bolted axial joint	螺栓轴向连接
bolted centre line joint	螺栓沿中心线连接
bolted joint	螺栓接合
bolting pad	螺栓垫
Bombardier Aerospace	（加拿大）庞巴迪宇航公司
Bombardier global express	庞巴迪全球快车（公务机）
Bombardier regional jet	庞巴迪支线喷气机
bonded logistic centre	保税物流中心
bonded warehouse	保税仓库
bonding by diffusion	扩散连接
book of reference	参考书
booster	增压压气机
booster aero parameter	增压压气机气动参数
booster and shaft design	增压压气机和机轴的设计
booster aspect ratio	增压压气机叶片展弦比
booster bleed valve actuator	增压压气机放气阀传动杆
booster containment	增压压气机包容性
booster cfficiency	增压压气机效率
booster engine	助推发动机
booster exit	增压压气机排气
booster inlet	增压压气机进气
booster outlet guide vanes	增压压气机排气导叶
booster pump	增压泵

booster rotor	增压压气机转子
booster spool tooth	增压压气机鼓齿
booster stage	增压级
booster stage bleed valve	增压器级间放气阀
bore	（发动机)孔洞
borescope	孔探仪
borescope access plugs	孔探仪插入口
borescope image	孔探图像
borescope inspection	用孔探镜检查
boresight error	瞄准误差
boron-fibre-reinforced composite	硼纤维增强的复合材料
bottle-fired indicator	灭火瓶指示器
bottom dead centre	基准线
bottom of loop	最低耗油点
bought-out finished materials	成品采购
bought-out parts	外购零件
boundary layer	边界层
boundary layer bleed	附面层泄除
boundary layer diverter	附面层隔道
boundary layer diverter system	附面层隔道系统
boundary layer drag	附面层阻力
boundary layer separation	边界层分离
boundary vorticity flux	边界涡量流
bow stator airfoils	弯曲静子翼型
bowed and swept cascade	弯掠叶栅
bowed stator	弯静叶
bracket	托架
brake cylinder	制动筒
brake drum	制动鼓
brake horse power	制动马力
brake hydraulic system	液压刹车系统
brake wear	刹车磨损
braking test	刹车试验
Brayton cycle	布莱顿循环

Brayton cycle performance	布莱顿循环性能
break away	克服静摩擦
break release weight	(飞机)最大允许起飞重量
breakdown voltage	击穿电压
break-even	(收支)平衡
break-even point	(收支)平衡点
breakpoint save	断点保存
breakthrough affordability	完全买得起
breakthrough technology	突破性技术
breather	油气分离器
breech cap	燃爆筒盖
Breguet range	布拉奎航程(航程的简化计算方法)
bribes for business	商业贿赂
Bridgeman casting process	布里奇曼铸造工艺
Brinell hardness	布氏硬度
Bristol Engine Division	(英国)布里斯托尔发动机部
Bristol Siddeley Engines	(英国)布里斯托-西德利发动机公司
Bristol Siddeley Engines Ltd. (now Rolls-Royce)	布里托尔赛德利发动机有限公司(现属罗罗公司)
British Aerospace Engineering Company	英国航空工程公司
British Aerospace Systems	(英国)英宇航公司
British Aircraft Corporation	英国飞机公司
British Civil Airworthiness Requirements	英国民用航空适航规定
British standard specification	英国标准化规范
British standard(s)	英国标准
British Standards Institution	英国标准化协会
British thermal unit	英国热量单位
brittle fracture	脆性断裂
broach	拉刀
broaching machine	拉床
broadband shock noise	宽带激波噪声
brush seal	刷式封严
brushgear	电刷
bucket door	铲斗式门

bucket specific fuel consumption	最佳耗油率
buffer air balance tubes	缓冲空气平衡管
buffer air heat exchanger	缓冲空气热交换器
buffer air manifold	缓冲空气总管
buffer air pressure sensor	缓冲空气压力传感器
buffer air valve	缓冲气阀
buffer cooler	缓冲冷却器
buffer stage bleed valve	缓冲级间气阀
buffer stock	缓冲库存
build	组装
build in quality	无缺陷
build package	整机组件包
build package integration	制造件集成
build rams	装配用升降机
build time	装配时间
build to order	按单订制
build-to-print	按图纸制造加工
build-up	组装好的发动机
built up propeller	组合螺旋桨
built-in relief valve	自带安全阀
built-in self-inspecting	机内自检
built-in test equipment	内置试验设备
bulk meter	容积流量计
bulk release	大量出图
bulkhead	舱壁板,隔离挡板
bump test	碰撞试验
bumpy flight	不平稳飞行
Bundesverband der Deutschen Luft und RaumfahrtIndustrie E. V.	德国航空宇航工业联合会
bunker surcharge	燃油附加费
Bureau of Transportation Statistics	(美国)运输统计局
burn off problem of lip	唇口烧坏问题
burner	燃烧器
burner feed arm	燃烧器输油臂

burner outlet temperature	燃烧室出口温度
burner pressure	火焰筒压力
burner rig test	台架燃气模拟
burner staging valve	分级燃烧阀
burning lean	贫油燃烧
burning rich	富油燃烧
burning ring	燃烧环
burrs	毛刺
burst of high energy parts through engine casing	高能部件破裂穿透发动机机匣
bus or transformer coupling	总线耦合或变压器耦合方式
Busemann inlet	布斯曼进气道
business activity	商务活动
business aircraft community	公务机行业
business case	商业可行性
business continuity plan	业务持续计划
business credit	商业贷款
business environment risk intelligence	商务环境风险情报
business ethics	商业道德
business evaluation review	商务评审会
business infrastructure	业务基础
business jet	公务机
business metrics	业务指标
business performance	经营成绩
business perspective	业务前景
business planning infrastructure	业务计划组织结构
business process	业务流程
business process model	业务流程模型
business process restructuring	重建业务流程
business requirements document	商务要求文件
business to business	企业对企业
business to customer	企业对消费者
bussing technique	总线技术
buy to fly ratio	原料与成品质量比
buyer furnished equipment	买方供货设备

buy-in module	支持模块
buzz noise	进场噪声
buzz-saw noise	圆盘锯噪声
bypass	外涵
bypass afterburner	涵道加力燃烧室
bypass duct	外涵道
bypass duct pressure loss	外涵道压力损失
bypass efficiency	外涵道效率
bypass engine	带外涵道的发动机
bypass pressure ratio	外涵增压比
bypass ratio	涵道比
bypass turbo-jet	涵道式涡轮喷气发动机
by-product	副产品

C duct reverser	C状外涵道反推装置
C/SiC composite	碳化硅陶瓷基复合材料
C - 17 Globemaster	C - 17 环球霸王运输机
cabin air	机舱空气
cabin air bleed	机舱放气
cabin conditioning system	座舱空调系统
cabin crew cost	机舱乘务组成本
cabin environmental control system	客舱环境控制系统
cabin noise	座舱噪声
cabin surveillance system	座舱监视系统
cabin voice recorder	座舱通话记录仪
cabotage	国内航空权
calibrated air speed	校准空速
calibrated airspeed	修正表速
calibrated engine	标定后的发动机
calibrated thrust	校准推力
calibration	校准

calibration record	校准记录
calibration standard	校准标准
calibration test	校准试验
called-up share capital	已催缴股本
calorific value	热值
camshaft	凸轮轴
camtrack	凸轮轨道
can（downstream of mixing plane）	共用喷气管（气体混合面后部）
can combustor	筒形燃烧室
Canada Regional Jet	庞巴迪支线客机（庞巴迪公司，加拿大）
Canadian Aviation Regulation(s)	加拿大航空规定
canard configuration	（飞机的）鸭式布局
Canberra	（英国）堪培拉飞机
Canberra	堪培拉–轰炸机（洛克希德–马丁公司，美国）
cancellation rate	（航班）取消率
can-do attitude	尽力做的态度
cannular combustion chamber	筒环形燃烧室
cannular combustor	环管燃烧室
cantilevered support	悬壁支承
capability build-up	能力的培养
capacitance storing energy type	电容储能式
capacity	流通
cap-and-trade system	（美国企业提出的减排）总量控制与配额交易
Capetown Convention	开普敦公约
capillary drilling	微孔钻削
capillary tube	毛细管
capital and liability ratio	资本负债比率
capital assets account	资本资产账户
capital bond	资本债券
capital cost	资本成本
capital expenditure	资本支出
capital reorganization	资本重组

Capstone	(美国)凯普斯通(微型燃机)公司
captured area	进气面积
Caravelle	快帆-客机(南方飞机公司,法国)
carbon and carbon complex material	碳碳复合材料
carbon constraints	限碳规定
carbon cycle	碳循环
carbon dioxide	二氧化碳
carbon dioxide emissions	二氧化碳排放
carbon dioxide tax	二氧化碳税
carbon emission	碳排放
carbon emissions trading	碳排放贸易
carbon fiber	碳纤维
carbon fiber wing	碳纤维机翼
carbon fiber-reinforced plastic	碳纤维增强塑料
carbon fiber-reinforced polymer	碳纤维增强聚合物
carbon fibre buffer bearing	碳纤维减震轴承
carbon fibre composite	碳纤维复合材料
carbon fibre reinforced aluminium	碳纤维铝合金层板
carbon fund	碳基金
carbon graphite seal	碳石墨封严
carbon monoxide	一氧化碳
carbon offset	碳补偿贸易
carbon seal	碳封严
carbon steel	碳钢
carbon trading	碳(配额)交易
carbon/carbon ceramic	碳碳陶瓷
carbon-fibre-reinforced composite	碳纤维增强的复合材料
carbon-free biofuel	无碳生物燃料
carburetor	汽化器
carburetor air heating	汽化器进气加温
carburization	渗碳
carburizing	渗碳
carcase vibration survey	机匣振动检测
carcinogens，mutagens and reproductive toxics	致癌物、诱变剂和可再生毒气

cargo freighter	货机
cargo hoist system	货运提升系统
Carlton	(美国铸造和机加工专门化公司)卡尔顿
Carlton Forge Works	(美国)卡尔顿锻造公司
carpet milling	毯式磨
carriage and insurance paid	运费加保险付至
carrier version	舰载型
cartridge firing unit	火药点火器
cartridge starter	火药起动机
cartridge turbo-starter	火药涡轮起动机
cascade	叶栅
cascade reverser	叶栅反推装置
cascade solidity	叶栅稠度
cascade thrust reverser	叶栅式反推装置
cascade vane	叶栅
Casebank Spotlight	自主学习故障隔离技术(普惠)
case-turbine exhaust	涡轮排气机匣
cash before delivery	交货前付款
cash dividends	现金股利
cash flow analysis	现金流量分析
cash flow cumulated	累计现金流量
cash flow statement	现金流表
cash funding	现金融资
cash on delivery	货到付款
cash operation cost	现金运营成本
cash squeeze	现金周转困难
cash with order	订货时付款
casing containment test	机匣包容试验
casing treatment	机匣处理
cast aluminium	铝铸件
cast blade ring	铸造的叶片环
cast floatwall segment	铸造浮壁段
cast head	上冠铸件

cast heatshield C1023	材料为 C1023 的铸造隔热板
cast in segment	扇形铸件
cast Inco 718 outlet guide vanes structure	材料为 Inco 718 的铸造排气导叶
cast superalloy	铸造高温合金
cast surface	铸造表面
casting	铸件
catalog price	标价
catalogue price	标价
Catalyst	"催化剂"(新型涡桨发动机,GE)
catalytic combustor	催化燃烧室
catalytic ignition	催化点火
category 1	第一类
Caterpillar	卡特彼勒公司
cathode ray tube	阴极射线管
cathodic vacuum etching	阴极真空蚀刻
cause & effect analysis	因果分析
cavitation	气穴
cavitation erosion	气蚀
cavity flow	空腔流
cavity flow pattern	空腔气流流动方式
cavity flows design	空腔流设计
"C" duct	"C"形短舱
center of pressure	压力中心
Central AeroHydrodynamic Institute	(俄罗斯)茹科夫斯基中央空气流体力学研究院
central aircraft information maintenance system	飞机信息维护系统中心
central drive device	中央传动装置
Central Flow Management Unit	中央流量管理中心
Central Institute of Aviation Motors	(俄罗斯)中央航空发动机研究院
Central Intelligence Agency	中情局
central processing unit	中央处理器
central sprue	中心注口
central warning panel	中央警告板
centre of gravity	重心

centre of gravity datum	重心基准
centre of gravity envelope	重心范围
centre of pressure	压力中心
centre sump	中央机油箱
centre wing section	中翼
centre wing tank	中翼油箱
centricast jethete (steel) split front casing	材料为合金钢的前部机匣
centrifugal atomizer	离心喷雾器
centrifugal blade load	叶片离心载荷
centrifugal breather	离心式油气分离器
centrifugal compressor	离心压气机
centrifugal compressor diffuser	离心压气机扩压器
centrifugal flow	离心气流
centrifugal friction clutch	离心式摩擦离合器
centrifugal impeller	离心压气机叶轮
centrifugal load	离心载荷
centrifugal mechanical tacho generator	机械离心式转速传感器
centrifugal pump	离心泵
Centro Tecnico Aeroespacial	巴西国家适航局
Ceramic Coating Center SA	（MTU 与 SAFRAN 合资）陶瓷涂层中心股份公司
ceramic composite materials	陶瓷复合材料
ceramic heat shielding material	陶瓷防热材料
ceramic insulator	陶瓷绝缘子
ceramic matrix composite	陶瓷复合材料
ceramic matrix composite analyzer	陶瓷基复合材料分析器
ceramic mould	陶瓷模
ceramic turbine blades	陶瓷涡轮叶片
ccrtificatc of conformity	合格证书
certificate of qualification	资质证书
certification/registration bodies	认证注册机构
certification maintenance requirement	审定维护要求
certification of maintenance	维修证明
certification process	取试程序

certification readiness review	适航审核
certification report or document	适航报告或文件
certification requirement	适航审定要求
certification test request	适航试验请求
certified emission reduction	核证的减排量
certified engine	已取证的发动机
certified engine definition	取证发动机的定义
certified financial statement	经会计师证明的财务报表
certified material test report	经认证的材料试验报告
certified release inspector	有资质的发证监察员
certified release inspector	有资质的发货监察员
Cessna	塞斯纳-通用飞机(达索公司,法国)
Cessna Aircraft Company	(美国)塞斯纳飞机公司
Challenger	挑战者-公务机(庞巴迪公司,加拿大)
chamber effect	舱效应
chamfer	倒角
chamfered to clear under bolt head	螺栓头下的倒角
change control review	变更控制审核
change request	变更要求
changed product ruling	更改产品规定
channel flow with variable mass flow rate	变流量管流
characteristic operation point	特征工作点
charging coefficient	充填系数
charpy impact strength	简支梁冲击强度
check bolt	防松螺栓
check nut	锁紧螺帽
check test	检验试车
chemical safety report	化学品安全性报告
chemical vapor deposition	化学气相沉积
chemical vapor deposition furnace	化学气相沉积炉
Cheng cycle	程氏循环
Chernyshev Mechanical Engineering Enterprise	(俄罗斯)切尔内舍夫机器制造厂
chevron	锯齿形喷口

chevron nozzle	锯齿型尾喷口
chief controls engineer	负责控制的总师
chief design engineer	总设计师
chief performance engineer	性能总师
chief validation engineer	负责实验的总师
China Aerospace Establishment	中国航空研究院
China Aerospace Science and Technology Corporation	中国航天科技集团公司
China Aerospace Science and Industry Corporation Limited	中国航天科工集团有限公司
China Aviation Industry Corporation Ⅰ	中国航空工业第一集团公司
China Aviation Industry Corporation Ⅱ	中国航空工业第二集团公司
China Aviation Suppliers Import and Export Group Corporation	中国航空器材进出口集团公司
china aviation weekly index	中国航空每周指数
China Civil Aviation Regulations	中国民用航空规章
China Council for the Promotion of International Trade	中国国际贸易促进委员会
China Gas Turbine Establishment	中国燃气涡轮研究院
China South Industries Group Co.，Ltd.	中国兵器装备集团有限公司
China North Industries Group Corporation Limited	中国兵器工业集团有限公司
China-Russia Commercial Aircraft International Co.，Ltd.	中俄国际商用飞机有限责任公司
chip detector	检屑器
chipping	碎屑
chlorofluorocarbon	氟氯碳化物
chocking spring	抗震弹簧
choked technique	堵塞技术
choker	节风门
choker characteristics	节风门特性
choking	堵塞
chromic acid anodizing	铬酸阳极化
chromium steel	铬钢

circuit	电路
circuit drawing	线路图
circuit load	线路负荷
circular head grooved bolt	圆头槽螺栓
circular triangle	圆弧三角形
circular truncated cone	截圆锥
circumferential dovetail	环向燕尾槽
circumferential forces	周向力
circumferential total pressure distortion sensitivity	周向总压畸变灵敏度
Citation	奖状-公务机(达索公司,法国)
civil aero engines	民用航空发动机
Civil Aeronautics Board	(美国)民用航空委员会
civil aerospace	民用航空
civil airway	民用航路
Civil Aviation Administration of China	中国民航局(原名)
Civil Aviation Authority	英国民航局
Civil Aviation Authority of Singapore	新加坡民航局
Civil Aviation Research and Development	民用航空研发
Civil Aviation Safety Authority	(澳大利亚)民航安全局
Civil Aviation University of China	中国民航大学
civil penalty	民事罚款
clamping arrangement	紧固装置
clamshell-type deflector door	贝壳形折流门
clapper	阻尼凸台
clash detection	冲突检查
classification of change	更改的分类
classification of the design change document	设计变更文件的分类
classified balance sheet	分类资产负债表
classified parts	分类零件
clay crucible	黏土坩埚
clean aircraft	环保型飞机
Clean Development Mechanism	(京都协议减排)清洁发展机制
Clean Sky	(欧盟)清洁天空研究计划

Clean Sky Joint Technology Initiative	(欧盟)清洁天空联合技术计划
Clean Sky Joint Technology Initiative	(欧盟资助的)清洁天空联合技术计划
clearance for bolts	安装螺栓的空间
clearances	间隙
clearway	净空道
clenched bolt	紧箍螺栓
climate change	气候变化
climb	爬升
climb rating	爬升功率
climb-to-altitude performance	爬升性能
clinometer	倾斜仪
clip stacks	卡箍
clock and pulse counter	时钟和脉冲计数器
clock wise	顺时针
close in combat	近距作战
close woven wire cloth	密集编织金属丝布
closed cycle	闭式循环
closed loop control	闭环控制
close-die-forging	闭模锻造
cloudburst treatment	喷丸处理
clustered vanes	集束导向叶片
coal field gas	煤层气
coal-bed gas	煤层气
coal-seam gas	煤层气
Coanda-effect	(圆盘概念飞机)柯恩达效应
coarse strainer	粗滤网
coating	涂层
coating shedding	涂层脱落
coating vapor deposition	气相沉积涂层
coaxial cable	同轴电缆
coaxial connector	同轴电缆接头套管
coaxial nozzle	同轴心喷管
coaxial rotor helicopter	共轴双旋翼直升机

cobalt based alloy	钴基合金
cobalt-chromium dispersion layer	钴-铬扩散层
cobonded	共胶接
cockpit display unit	驾驶舱显示装置
cockpit indications	驾驶舱指示
cockpit pressure control switch	驾驶舱压力调节开关
cockpit temperature regulating system	驾驶舱温度调节系统
cockpit voice recorder	驾驶舱录音机
cockpit with video graphic displays	驾驶舱配有视频图像显示器
code-share agreement	(航空公司之间)代码共享协议
co-developing	联合开发
coefficient of drag	阻力系数
coefficient of lift	升力系数
coefficient of viscosity	黏性系数
coiled pipe cooler	盘管冷却器
coking	焦化
cold drawing	冷拉
cold forging	冷锻
cold forming	冷成型
cold junction temperature	冷结合温度
cold pressing	冷压
cold rolling	冷轧
cold run	冷运转
cold start	冷启动
cold-end drive	冷端驱动
collaborative optimization	协同优化
collecting information network	信息收集网络
collector tray	集油槽
color pyrometer	比色高温计
columnar crystal structure	柱状晶体结构
combat air patrol	空中战斗巡逻
combat manoeuvres	空战演习
combat search and rescue	战场搜寻和救援
combined atomizing nozzle	混合式雾化喷嘴

combined compressor	组合压气机
combined cooling blade	组合式冷却叶片
combined diesel engines and gas turbine	柴油机和燃气轮机联合循环
combined diesel engines or gas turbine	柴油机或燃气轮机联合循环
combined drive device	混合传动装置
combined engine	组合发动机
combined engine removal rate	综合换发率
combined environmental reliability test	组合环境可靠性试验
combined heat & power	热电联产
combined overspeed & shut-off solenoid	组合超速限制螺线管
combined turbine	复合涡轮机
combined-cycle power plant	联合循环发电厂
combustion chamber	燃烧室
combustion chamber and tile	火焰筒和内衬
combustion chamber discharge nozzle	燃烧室燃气导管
combustion chamber outer casing	燃烧室外部机匣
combustion effectiveness	燃烧效果
combustion efficiency	燃烧效率
combustion gas dynamics	燃烧气体动力学
combustion instability	燃烧不稳定性
combustion intensity	燃烧室容热强度
combustion line	燃烧室衬套
combustion product	燃烧产物
combustion simulation criteria	燃烧模拟准则
combustion stability	燃烧稳定性
combustor	燃烧室
combustor casing	燃烧室机匣
combustor dilution	燃烧室掺混
combustor effusion cooling	燃烧室多孔冷却
combustor exit temperature	燃烧室出口温度
combustor flow fields	燃烧室流场
combustor liner	火焰筒
combustor outlet temperature	燃烧室出口温度
combustor pressure loss	燃烧室压力损失

combustor reference velocity	燃烧室平均流速
combustor rig test	燃烧室部件试验
combustor test	燃烧室试验
combustor tile	燃烧室壁板
combustor total pressure	燃烧室总压
combustor total pressure loss coefficient	燃烧室总压损失系数
combustor with rotary atomizer	甩油盘燃烧室
Comet	彗星-客机(德哈维兰公司,英国)
commerce at light speed	光速商务
commerce control list	(美国)商务管制清单
commercial air transport	商用航空运输
commercial and government entity	商业和政府权益
commercial engine business	民用发动机业务
commercial invoice shipping	商业货运发票
commercial maintenance	商业维修
commercial modular aero-propulsion system simulation	商用模块化航空推进系统模拟
commercial supply specification	商务供货规范
commercial terms agreed	已达成商务条款
commercial transport	民用飞机
Commins	(美国)康明斯公司
Commission of Science，Technology and Industry for National Defence	国防科学技术工业委员会
commissioned overcheck policy	请第三方复检政策
Committee on Aviation Environmental Protection	航空环境保护委员会
commodity leader	产品采购领导人
commodity risk	商品风险
common core	通用核心机
common data network	通用数据网络
common extensible cryogenic engine	(美国国家航空航天局)通用可扩展的低温发动机
common external tariff	统一对外海关税率
common fuselage	通用机身

common justification	通用解释
common raw material	通用原材料
common stock equity	普通股权益
commonality advantage	通用性的优势
commonality with underwing application	与翼下吊装具通用性
commons defence committee	(英国)下院国防委员会
commutator end plate	整流器端板
commuter aircraft	支线飞机
commuter airline	短途航空公司
compact airborne early warning	紧凑型空中预警飞机
compact and less weight engine	紧凑重量更轻的发动机
compact disk-read only memory	光盘读入内存系统
company calibration procedure	公司校准流程
company materials engineering	公司的材料技术(罗罗)
company overview	公司简介
company quality acceptance standard	质量验收标准(罗罗)
company quality control procedure	质量控制程序(罗罗)
comparative vacuum monitoring	(无损探伤)比较真空监测
compatible air	匹配的空气
Compbell diagram	(振动分析)坎贝尔图
competency	胜任特征
competency model	胜任特征模型
competent and wealthy competitors	有能力和实力的竞争者
competent person	合资格的人
competitive assessment	竞争性分析
competitive price	有竞争力的价格
competitiveness increase	增强竞争力
complex	复杂零件
complex formulae	复杂的公式
complex shaped cooling holes and ducts	复杂的成形冷却孔和导管
compliance check list	一致性检查清单
compliance testing	符合性试验
component approval package	部件核准包
component care process	部件处理流程

component cost worksheet	部件成本工作单
component deactivating	解除部件功能
component definition group	部件定义组
component degreasing	部件去油污
component derusting	部件除锈
component development plan	部件开发计划
component development programme	部件开发规划
component engineering procedure	部件设计流程
Component Fan	(GE 和斯奈克玛公司合资)复合材料风扇叶片公司
Component Improvement Program	(美国)发动机部件改进计划
component maintenance manual	部件维护手册
component manufacturing technique	部件制造技术
component matching	部件匹配
component modal impact test	零部件模态分析实验
component performance	部件性能
component proving	部件审核过程
component refurbishment	部件翻新
component replacement cost	部件拆换费用
component retirement life limitation	部件退役寿命限制
component shop manual	部件工装手册
component specific rationalised quality standard	部件特定质量优化标准
component tracking filter	部件跟踪滤波器
component ultimate life limitation	部件最大寿命限制
components offered	供货部件
composite	复合材料
composite construction	复合材料结构
composite die	组合模
composite engine	组合发动机
composite erosion	复合材料侵蚀
composite erosion control	复合材料侵蚀控制
composite material	复合材料
composite material fiber placement	复合材料丝束铺放
composite material helicopter	复合材料直升机

compound leaned blade	复合倾斜叶片
compound turbine	复式涡轮机
compressed natural gas	压缩天然气
compressible flow	可压缩流
compressible flow curve	可压缩流曲线
compression	压缩
compression ignition	压缩点火
compression ignition engine	压燃式发动机
compression pressure	压缩压力
compression ramp	压缩斜板
compression ratio	压缩比
compression shock	压缩激波
compression work	压气机功
compressive strength	压缩强度
compressor	压气机
compressor air flow	压气机空气流量
compressor and turbine rotor thermal survey	压气机和涡轮转子温度监测
compressor and turbine vibration survey	压气机和涡轮振动监测
compressor blade flutter measurement	压气机叶片颤振测量
compressor characteristics	压气机特性
compressor casing	压气机机匣
compressor delivery pressure	压气机出口压力
compressor disc	压气机盘
compressor discharge temperature	压气机出口温度
compressor discharge air	压气机排气
compressor discharge pressure	压气机出口压力
compressor drums	压气机鼓筒
compressor efficiency	压气机效率
compressor element stage	压气机基元级
compressor exit	压气机出口
compressor exit diffuser	压气机出口扩压器
compressor flutter	压气机颤振
compressor handling bleed	压气机放气阀
compressor handling bleed valve	压气机排气控制阀门

compressor inlet	压气机进气
compressor intermediate case	压气机中介机匣
compressor interstage bleed	压气机级间引气
compressor locked stall	压气机锁定失速
compressor map	压气机特性图
compressor operating point	压气机工作点
compressor pressure ratio	压气机增压比
compressor regulation	压气机调节
compressor rig test	压气机部件试验
compressor rotating stall	压气机旋转失速
compressor rotor	压气机转子
compressor rotor blade	压气机转子叶片
compressor secondary characteristic	压气机二级特性图
compressor stability margin	压气机的稳定裕度
compressor stage	压气机级
compressor stall	压气机失速
compressor stall measurement	压气机失速测量
compressor stator vane	压气机整流叶片
compressor surge	压气机喘振
compressor surge control	压缩器喘振控制
compressor surge margin	压气机喘振裕度
compressor tertiary characteristic	压气机三级特性图
compressor test	压气机试验
compressor tip clearance	压气机叶尖间隙
compressor variable inlet guide vane	压气机可调进口导叶
compressor variable inlet guide vane levers	压气机可调进口导叶摇臂
compressor variable stator vane	压气机可调静子叶片
compressor working fluid	压气机工质
compressor working line	压气机工作线
computational fluid dynamics	计算流体力学
computational fluid dynamics of engines	发动机计算流体力学
computational fluid dynamics program	计算流体动力学程序
computer added engineering	计算机辅助工程
computer aided design	计算机辅助设计

computer aided engineering	计算机辅助工程
computer aided innovation	计算机辅助创新
computer aided manufacture	计算机辅助制造
computer aided plan	计算机辅助计划
computer aided quality control	计算机辅助质量控制
computer aided three-dimensional interactive application	计算机辅助三维交互式设计软件
computer assisted passenger prescreening system	计算机辅助乘客预先筛检系统
computer based training	基于计算机的培训
computer integrated manufacturing system	计算机集成制造系统
computer numerical control 5-axis machining center	数控五坐标加工中心
computer numerical control lathes	数控车床
computer numerical control machined parts	数控加工零件
computer numerical control machining	数控加工
computer numerical control vertical grinder	数控立式磨床
computer simulation analysis and evaluation system	计算机模拟分析与评估系统
computer supported cooperative work	计算机协同工作
computer-aided process planning	计算机辅助工艺规划
concave skin	凹面蒙皮
concept demonstration aircraft	（F135）方案验证机
concept design	概念设计
concept design and analysis	方案设计和分析
concept review	方案审核
concession	（对零件超差的）特许
concession label	特许标识
concession note	特许说明
concessions on engine	发动机优惠价格
Concorde	协和号超声速客机
concurrent engineering	并行工程
concurrent subspace optimization	并行子空间优化
condensate pump	凝水泵

condensation	冷凝
condensor	冷凝器
con-di fan nozzle	收敛-扩散式风扇喷管
con-di intake	收敛-扩散式进气道
condition based maintenance	视情维护
condition lever angle	油门控制位置
condition of supply	供货条件
conditioned air ground system connection	空调系统地面接头
Condor	秃鹰-无人机
conduction	传导
conductive slip	导电滑移
conductive slip ring	导电滑环
configuration	构型
configuration definition and control	方案的定义和控制
configuration management system	构型管理系统
conformance control feature	一致性控制特征
conic coupling	圆弧篦齿联结
conical bearing	圆锥轴承
conical shaping	锥面造型
connected simulated altitude test	连接式模拟高空试验
connecting features	联结特性
conservation of mass	质量守恒
conservation of momentum	动量守恒
consistency over-check	一致性复检
consolidate business review	重组业务审查
consolidated balance sheet	合并资产负债表
consolidated financial statement	合并财务报表
consolidating gain	巩固收益
constant cycle exchange rate	恒定循环的敏感性
constant fuel flow regulator	恒供油量调节器
constant pitch	不变桨距
constant pressure drop valve	等压差活门
constant pressure valve	定压活门
constant specific thrust	恒定特征推力

constant speed drive	恒速驱动
constant speed unit	恒速装置
constant temperature control valve	恒温控制活门
constant volume combustion	定容燃烧室
constant volume combustion	等容燃烧
constant-stroke injection pump	定行程喷油泵
constant-volume-combustion-cycle engine	定容燃烧循环发动机
Construcciones Aeronáuticas S. A	西班牙航空工程公司
Construction And Detailing Design System	构型和详细设计系统（设计软件）
consulting strategy	战略咨询
consumable and replenishment spares	消耗和补充性备件
containment design	包容机匣设计
containment device	包容装置
containment ring	包容环
contaminated fuel	污染燃油
contamination	污染
contamination analysis	污染物分析
contamination of bleed air	引气污染
content detection limit	含量检测范围
contingency special airlift mission	（美国）紧急特殊空运任务
continued airworthiness	持续适航
continued airworthiness assessment methodologies	持续适航评估方法
continued airworthiness of the engine	发动机持续适航
continuous casting	连续铸造
continuous curvature	连续曲率
continuous descent approach	连续下降进近
continuous groove interstage	级间连续槽
continuous lower energy, emissions and noise	持续低能耗,低排放,低噪声
contra rotating propeller	对转螺旋桨
contract assurance instruction	合同保证规定
Contract Review Board	合同审核委员会
contractual service agreement	合同服务协议
contra-flow regenerator	逆流回热器

contrails	尾迹
contra-rotating fan	对转风扇
contra-rotating fan-concept	对转风扇方案
contra-rotating intershaft bearing	对转中介轴承
contra-rotating shaft	双轴对转
contra-rotating turbine	对转涡轮
control and audit of specification system	特定系统的控制和审核
control chart	控制图
control element	控制元件
control of substances hazardous to health （UK）	(英国)有害物质控制
control pressure ratio engine	可控压比发动机
control system	控制系统
control system aircraft interface document	控制系统飞机界面文件
control system stability	控制系统稳定性
control valve	控制阀
controllable engine removal rate	可控制换发率
controllable pitch propeller	可控桨距螺旋桨
controllable reversible pitch	可控逆桨距
controlled diffusion aerofoil	防扩散气动叶型
controlled diffusion airfoil	防扩散叶型
controlled diffusion airfoil blading	控制气流分离叶片
controlled features	控制特性
controlled vortex design	可控涡设计
convection cooling blade	对流冷却叶片
convective cooling	对流冷却
conventional aircraft design	常规飞机设计
conventional profile	常规翼型
conventional take-off and landing	常规起飞着陆
conventional turbofans	普通涡扇发动机
convergent ducts	收敛涵道
convergent nozzle	收敛喷管
convergent-divergent nozzle	收敛-扩张喷管
convex skin	凸面蒙皮

Conway	康威发动机（罗罗公司，英国）
cool by-pass airflow	低温外涵空气流
coolant pump	冷却泵
cooled shrouds	带冷却的叶冠
cooling air control device	冷却空气控制机构
cooling air pre-swirl nozzles	冷却空气预旋喷嘴
cooling and sealing flow	冷却和封闭气流
cooling effectiveness of turbine blade	涡轮叶片冷却效果
cooling flow through the clearance between tenon and mortise	榫头装配间隙吹风冷却
cooling medium	冷却介质
cooling plate	冷却板
cooling tower	冷却塔
cooling/mainstream flow interaction	冷却气流和主气流的相互作用
cooperation in government engine programs	政府发动机项目的合作
coordinate measuring machine	坐标测量仪
copilot	副驾驶员
cordite charge	火药桶装药
cordite starter	火药起动机
core competency	核心业务能力
core demonstrator test programme	核心验证机测试计划
core engine	核心机
core engine design and manufacture	核心机的设计和制造
core engine mounted	核心机吊装
core engine nozzle throat	核心机喷管喉道
core engine programme	核心机计划
core engine size	核心机尺寸
core engine test programme update	核心机试验计划现况
core engine washing	核心机清洗
core engine water ingestion testing	核心机吸水试验
core flowpath	内涵道
core ingestion	核心机异物吸入
core location dowel	型芯定位销
core location moulding box	型芯位置铸模盒

core stubshaft	锥形短轴
core-driven fan stage	核心机驱动风扇级
core-fan mixing chutes	核心机和风扇气流混合道
core-mount	核心机吊装
corner stall	角失速
corner velocity	(战斗机)最大转弯速度
corporate capital	公司资本
corporate responsibility	企业责任
corporate social responsibility	企业社会责任
corporate spend	公司开销
CorporateCare program	公务机发动机包修项目(罗罗)
correct part number	正确的部件标号
correct process routing	正确处理过程
corrected parameter	换算参数
corrected rotating speed	修正转速
correction unit area	校正单位面积
corrective action report	纠正措施报告
corrective maintenance	改正性维护
corrosion	腐蚀
corrosion crack	腐蚀性裂纹
corrosion damage	腐蚀损伤
corrosion fatigue	腐蚀疲劳
corrosion inhibiting	防锈密封
corrosion protection layer	腐蚀保护层
corrosion resistant steel	耐蚀钢
corrosion test	腐蚀试验
corrosion treatment	腐蚀处理
corrugated internal mixer	波纹形内部混合器
corrugated joint	波纹形连接
corrugated strip	波纹板
Corsair	海盗-攻击机(沃特公司,美国)
Corvette	克尔维特-公务机(塞斯纳公司,美国)
cost account manager	成本核算经理

cost and freight	成本加运费
cost compliance	成本一致性
cost effective	低成本
cost effective solution	低成本解决方案
cost effective technology	低成本技术
cost estimate relation	成本估算法
cost index	维护费用系数
cost of capital	资金成本
cost of goods sold	产品销售成本
cost of non-quality	非质量成本
cost overrun	超支
cost per available seat mile	标准里程成本
cost reimbursable contract	成本偿还合同
cost，insurance and freight	成本、保险加运费
cost-cutting	降低成本
costed technical programme	项目工程成本
cost-effective working	经济有效的工作
costly equipment	昂贵设备
costly redesign	高价重新设计费
cost-of-living index	生活费用指数
cost-optimized production	优化成本生产
cost-per-flight-hour agreement	按飞行小时计费协议
cost-plus-fixed-fee contract	不付超支的固定合同
costs of manufacture	制造成本
costs of production	生产成本
count rotating open rotor	对转桨扇开式转子发动机
counter air	（禁止敌方飞行器升空的）制空行动
counter clockwise	逆时针
counter rotating integrated shrouded propfan	对转整体涵道式桨扇发动机（德国 MTU 公司）
counterbore	反向钻孔
counterclockwise readings	逆时针读数
counterparty credit risk	交易对手信贷风险
counter-rotating turbine	对转涡轮

countersunk head rivets	沉头铆钉
countersunk rivet	沉头铆钉
countervailing duties	反补贴税
countervane	导向叶片
coupling	联轴器
coupling assembly	联轴节组件
coupling bolt	联结螺栓
coupling shaft	联轴器轴
course	航向
course angle	航向角
course indicator	航道指示器
cow anti-ice air control value	短舱防冰气控阀
cowl door	整流罩门
crack	裂纹
crack growth life	裂纹增长寿命
crack initiation	裂纹萌生
crack propagation	裂纹扩展
crank shaft-connecting rod mechanism	曲轴连杆机构
create customer solutions	为客户创造解决方案
creation of the new design definition document	编制新设计定义文件
credit crunch	信贷紧缩
credit rating	信用评级
credit standing	商业信誉
creep	蠕变
creep failure	蠕变失效
creep fatigue	蠕变疲劳
creep feed grinding	蠕动磨削
creep fracture	蠕变断裂
creep length	蠕变长度
creep life	蠕变寿命
creep relaxation	蠕变松弛
creep resistance	抗蠕变
creep rupture	蠕变断裂
creep stress	蠕变应力

creep time	蠕变时间
crew alert system	机组警报系统
crew altering	机组变更
crew altering system	机组人员变更系统
crew operation manual bulletin	机组操纵手册通告
critical deformation	临界变形
critical design review	关键性设计评审
critical hardening	临界淬火
critical icing condition	临界结冰状态
critical joints	关键接头
critical part	关键零部件
critical pressure	临界压力
critical propulsion components	关键推进系统部件
critical rotor speed	转子临界转速
critical speed	临界速度
critical speed modeshape	临界转速模态
critical supplier review	重要供应商审核
critical to quality	品质关键
cross check	交叉检验
cross engine debris criteria	侧向异物冲击的要求
cross keyed honeycomb seal	十字交叉蜂窝状封严篦齿
cross wind stability	侧风稳定性
cross-functional	跨部门运作
crosswind	侧风
crosswind inlet flow	侧风进气流
crosswind landing	侧风着陆
crosswind test	侧风试验
crown	焊缝顶部
cruise ceiling	巡航高度
cruise incidence angel	巡航机翼倾角
cruise rating	巡航功率
cruise specific fuel consumption	巡航耗油率
cruising altitude	巡航升限
cruising speed	巡航速度

cryogenic shutoff valve	低温截流阀
cryogenic test	低温测试
crystal boundary	晶界
crystal grain	晶粒
crystal grain-growth	晶粒生长
crystallisation zone	结晶区
crystal-type vibration transmitter	晶体式振动传感器
cube law	三次方关系
cumulative failures	累计失效次数
cumulative fatigue index	累计疲劳指数
cumulative margin	累积裕度
cupreous powder metallurgic material	粉末冶金铜基材料
currency adjustment factor	汇率调节附加费
current assets	流动资产
current certified definition	通用已认证定义
current engine option	选装原发动机方案
current International Civil Aviation Organization limit	现行国际民航组织规定值
current liability	流动负债
current market outlook	当前市场展望
curved dovetail foot	曲线燕尾榫头
curvic coupling	圆弧箟齿联轴器
curvic coupling joint to high pressure compressor	连接高压压气机的圆弧箟齿联轴器
curvic coupling to fan	连接风扇的圆弧箟齿联轴器
curvic coupling to fan stubshaft	连接风扇短轴的圆弧箟齿联轴器
curvic coupling to high pressure turbine	连接高压涡轮的圆弧箟齿联轴器
curvy gear grinder	弧端齿磨床
Custer channel wing	卡斯特沟槽机翼
customer bleed extraction	从客户定制气阀引气
customer bleeds	定制放气
customer change request	客户变更要求
customer concession	客户超差呈报
customer configuration file	客户构型档案

customer critical design review	有客户代表参加的关键设计评审
customer deck	用户性能手册
customer deck standard	用户性能标准手册
customer demand rate	客户需求周期
customer engagement	客户咨询
customer experience management	客户体验管理
customer facing business unit	面对客户的业务单位
customer focus	客户第一
customer focused	关注客户
customer information system	客户信息系统
customer lifetime value	客户生命期价值
Customer Management Point	客户管理点
customer relation management	客户关系管理
customer service and logistics	产品服务与支援
Customer Service Center Europe GmbH	(德国 MTU)客户服务中心欧洲公司
Customer Service Center，Ludwigsfelde	(德国 MTU)客户服务中心-路德维希斯菲尔德
customer support center	客户支援中心
customer survey	客户调查
customer working together team	与客户协同工作组
customer-designed	为客户而设计的
customized solutions	客户化方案
customs clearance	结关
cutaway view	剖面图
cutback	(降噪)减油门起飞
cut-off valve	断油活门
cutting edge industry	尖端工业
cutting edge of...	在……领先
cutting edge technologies	创新技术
cyborg insects	电子昆虫
cycle deck	发动机性能计算机模型
cycle design	发动机循环设计
cycle matching	循环匹配
cycles per second	每秒循环数

cyclic life sampling	寿命期抽样检查
cyclic testing	循环试验

daily utilisation	日利用率
damage	损坏
damage accumulation theory	累积损伤理论
damage resistance	抗损伤能力
damage tolerance criteria	损伤容限准则
damage tolerance design	损伤容限设计
damping	阻尼
damping effect	阻尼作用
damping mode	阻尼状态
damping sleeve	阻尼套筒
damping slice	阻尼片
Dart engines	达特发动机
dashpot throttle	阻尼油门
Dassault Aviation	（法国）达索公司
data acquisition computer	数据采集计算机
data acquisition unit	数据采集器
data control unit	数据控制器
data element identification	数据元识别
data entry plug	数据输入接口
data exchange	数据交换
data handling speed	数据处理速度
data model	数模
data storage unite	数据存储器
Data System and Solution	数据系统和解决方案公司（罗罗）
data transfer cartridge	数据转换卡
data transmission sheet	数据传输表
date code	生产日期码
datum	基准

datum condition	参考条件
de Haller number	德哈勒数
de Havilland Aircraft Ltd.	(英国)德哈维兰飞机公司
de Laval bulb root	拉瓦尔球形叶根
dead crank	重载起动
dead weight tonnage	载重吨位
dead-soft annealing	(镍合金在较高温度和较长时间)保证软化的退火
deaerator	油气分离器
de-aerator tray	油气分离器盘
debris of released blade	叶片断裂产生的碎片
debugging	调试
decel	减速
decel transient	减速瞬态
decent below glideslope	低于滑角的下降
decibel	分贝(噪声单位)
decision altitude	决断高度
decision height	决断高度
declaration of design and performance	设计和性能的取证申报
declared cyclic life	呈报循环寿命
declared safe cyclic life	取证申明安全循环寿命
decomposition temperature	热裂解温度
decompression valve	减压阀
decontamination and cleaning methods	除污和净化方法
dedicated capacity	专用产能
dedicated machine	专用设备
deep stall	严重失速
defect analysis	缺陷分析
defect parts	有缺陷部件
defective product	残次品
defects	缺陷
defects per unit	缺陷率
defence aerospace	防务航空业务
defence condition	防务条件

defence contract management	防务合同管理
defence contract management agency	防务合同管理局
Defence Industries Council	(英国)国防工业委员会
Defence Logistics Organisation	(英国)防务后勤机构
Defence Quality Assurance Group	(英国国防部)防务质量保证组
defence specification (UK)	英国防务规范
defence spending	国防费用
defence standard (UK)	英国国防标准
Defence Technology Centre	国防技术中心
Defense Advanced Research Projects Agency	(美国)国防预先研究计划局
Defense Security Cooperation Agence	(美国)国防安全协作局
defensive aids suite	辅助防御系统
deferred assets	递延资产
define, measure, analyse, improve, control	定义、度量、分析、改进、控制
definitive engineering model	确定的工程模型
deflagration wave	缓燃波
deflected thrust	偏转的推力
deflected thrust-vectoring	带倾角的推力矢量
deflection angle	偏转角
deflection limiter	挠度限制器
deflector door	折流门
deformation wear	变形磨损
degree of augmentation	加力比
degrees celsius	摄氏度(℃)
de-icing	除冰
delamination	分层
delay and cancellation	延误和取消
delay rate	(航班)延误率
delayed elasticity	延迟弹性
delayed fracture	延迟断裂
delayed yield	延迟屈服
delivered at frontier	边境交货
delivered duty paid	完税后交货
delivered duty unpaid	未完税交货

delivered ex quay	目的港码头交货
delivered ex ship	目的港船上交货
delivery test	交付试车
delta-wing	三角翼
delta-wing aircraft	三角翼飞机
Demag Delaval Industrial Turbomachinery AB	西门子在瑞典的工业燃机分公司
demand export license	要求出口许可证
demonstrator	验证机
demonstrator engine	发动机验证机
dense design	紧凑型设计
density ratio	密度比
dent	凹痕
dent in surface	表面凹陷
deoiler	油气分离器
Department of Defense	美国国防部
Department of Energy	美国能源部
Department of Trade and Industry，UK	英国贸工部
Department of Transportation	美国交通运输部
department procedure instruction	部门工作条例
depreciation	折旧
depth profiling	深度剖面分析
Derby engine division	(英国)达比发动机部
deregulation	解除管制
de-risking technology	低风险技术
Derwent engines	"德温特"发动机(罗罗)
Derwent process	德温特工程管理程序(罗罗)
desalination plan	海水淡化计划
design and make	设计并制造
design authority	主管设计部门
design by analysis	分析设计
design by rules	规则设计
design certificate	设计证书
design change approval form	设计更改批准表

design change document	设计更改文件
design change impact identification	设计更改影响的判明
design change proposal	设计变更建议
design characteristics impact	设计特征的影响
design definition issue statement	设计定义文件说明
design definition review and audit meeting	设计定义评审会
design failure mode effects analysis	设计故障模式和后果分析
design flaw	设计缺陷
design flow	设计流量
design for manufacture	可生产设计
design for manufacture & assembly	为制造和组装而设计
design for six Sigma	六个希格玛设计
design for test	可测试设计
design freeze	设计冻结
design from scratch	从头开始设计
design issue standard	设计文件标准
design life	设计寿命
design make build guide	设计产生制造指南
design methodology	设计方法
design organisation approval	设计机构认证
design organization exposition	设计组织说明
design organization handbook	设计组织手册
design point	设计点
design point of compressor	压气机设计点
design point performance	设计点性能
design precaution	设计防护
design range	设计航程
design requirement	设计要求
design responsible party company	负责设计的合作公司
design review	设计评审
design scheme issue note	设计出图说明
design sketch	设计草图
design specifications	设计技术规范
design standard	设计标准

design target	设计目标
design verification review	设计鉴定审核
designated engineering representative	(美国联邦航空局)委任工程代表
designated part warranted life	指定部件担保寿命
designated quality representative	特派质量代表
design-build teams	设计制造团队
designed inside-out	全面设计
designed modification design organisation representative	派驻改装设计机构的代表
desilvering	脱银
destroyer	驱逐舰
destructive physical analysis	破坏性物理分析
detachable splitter	组合式分离器
detectability	可检测性
detector	探测器
detonation	爆震
detonation wave	爆震波
Deutsche Zentrum fuer Luft-und Raumfahrt	德国航空航天研究院
Deutsche Zentrum fuer Luft-und Raumfahrt Köln	德国航空航天研究院科隆分部
Deutscher Kalibrier Dienst (German Calibration Service)	德国标定服务中心
develop existing supply chain process	现有供应链的开发过程
development and certification engine	开发与适航取证用发动机
development and innovation of safety management	安全管理的发展与创新
development center	开发中心
development material specification	材料开发规范(罗罗)
development program	开发计划
development rationalised process specification	合理化开发流程规定
development rationalised quality standard component	合理化开发部件质量标准
development rigs	研发用试验台
development risk	开发风险

development test	研制试验
deviation advanced engineering memorandum	先进工程超差备忘录
deviation angle	偏角
deviation defect request	超差请求
deviation permit	偏离许可
dewaxing	脱蜡
diagnose contactless thermal fault	非接触式热故障诊断
diagnostic health monitoring	诊断与状态监测
diaphragm	膜片
die cast	模铸
die casting	压铸
die casting process	压铸工艺
die forging	模锻
diesel engine	柴油发动机
diesel fuel	柴油
diethylene glycol	二甘醇
differential engine throttling	发动机差动节流
differential pressure	压差
differential pressure gauge	压差计
differential pressure indicator	压差指示器
differential pressure measurement	压差测量
difficult-to-cut materials	难加工材料
diffuser	扩压器
diffuser case	扩压器机匣
diffuser effectiveness	扩压器有效系数
diffuser of combustor	燃烧室扩压器
diffuser static pressure recovery coefficient	扩压器静压恢复系数
diffuser vane	扩压器导叶
diffusion bonded	扩散连接
diffusion coefficient	扩散系数
diffusion depth analysis	扩散深度分析
diffusion flame	扩散火焰
digital analog comparison	数模对比
digital design	数字设计

digital drawing	数字制图
digital electronic engine control	数字式电子化发动机控制
digital electronic flight control system	数字电子飞控系统
digital engine control unit	数字式发动机控制装置
Digital Enterprise Lean Manufacturing Interactive Application	(法国达索)数字化企业互动精益制造软件
digital factory	数字工厂
digital flight data recorder	数字式飞行数据记录器
digital head-up display	数字化平视显示器
digital indication	数字显示
digital mannequin	数字化人体模型
digital mockup	数字样机
digital part	数字零件
digital parts tracking	电子化的部件跟踪
digital pre-assembly	数字化预装配
digital product	数字产品
digital product definition	数字化产品设计
digital readout	数字显示
digital signal processing	数字信号处理
digital tachometer	数字转速仪
digital terrain elevation data	数字化地形特征数据
dihedral vanes	倾斜静叶
diluted stator combustor	掺混静子式燃烧室
dilution air holes	稀释空气孔
dilution hole	掺混孔
dilution inner port	掺混内端口
dilution zone	混合区
dilution zone	掺混区
dimensionless groups	无量纲参数组
direct current	直流电
direct current dynamo	直流电机
direct liability	直接负债
direct maintenance cost	直接维护成本
direct maintenance cost guarantee	直接维护成本保证

Direct Matrix Abstraction Program	（NASTRAN 用户编程语言）直接矩阵提取编程
direct operation cost	直接运营成本
direct part marking	直接部件标记
direct solidified	定向结晶
direction control unit	方向控制器
direction control valve	方向控制阀
Direction Générale de l'Aviation Civile	法国民航总局
directional crystal foundry	定向结晶铸造
directional pattern distortion	方向图畸变
directional property	方向性
directional solidification	定向凝固
directionally solidified blade	定向结晶叶片
directionally solidified turbine blade	定向结晶涡轮叶片
director general submarines	潜艇业务总经理
dirty fuel test	脏油试验
disc burst	盘破损
disc burst speed	轮盘破裂转速
disc burst test	轮盘破裂试验
disc chamfer	盘斜面
disc overspeed test	轮盘超转试验
disc rim speed	盘缘速度
disc strength calculation	轮盘强度计算
disc vibration	轮盘振动
disc's flange	盘缘
discharge coefficient	流量系数
discharge flow distortion	出口流场畸变
discharge gap	放电间隙
discharge noise	排气噪声
discharge nozzles	排放喷嘴
discharge pressure	流量压力
discharge swirl	出口旋流
discharge to atmosphere	排入大气
disciplined approach	有序的步骤

disc-oiled bearing	圆盘注油轴承
discoloration	变色
disconnection protection	断线保护
discontinuity	间断
discrete frequency generator	离散型频率发生器
discrete noise	离散噪声
dismantling and disassembly	拆卸和分解
dispersion coating	沉积涂层
displacement	位移
dissociation	离解
dissolved water	溶解水
distortion	扭曲变形
distortion assessment during life cycle	寿命期内畸变评定
distortion coefficient	畸变系数
distortion index	畸变指数
distortion pattern	畸变图谱
distortion screening parameter	畸变筛选参数
distortion tolerance	畸变容限
distortion transfer	畸变传递
distributed control system	分布式控制系统
distributed development	分布式开发
distribution channels	销售渠道
divergent ducts	扩散涵道
diverging isobars	分散等压线
dividend accumulation	累计红利
dividend from capital	资本股利
document numbering system	文件编号系统
dollarisation	美元化
dollarisation of supply chain	供应链美元化(在美元区发展供应链)
door opening system	开门系统
doppler flow transducer	多普勒流量传感器
double annular combustor	双环腔燃烧室
double bypass engine	双外涵道发动机
double casing	双层机匣

double circle arc	双圆弧形状
double circular section airframe	双圆截面机身
double diaphragm disc	双隔膜盘
double element thermocouple	双元热电偶
double ended stator	双端面导向器
double flow turbine	双流涡轮机
double fuel combustion system	双燃料燃烧系统
double perforate layer	双层孔板
double slotted flap	双开缝襟翼
double-decker plane	双层飞机
double-entry single-stage centrifugal turbo-jet	双面进气单级离心式涡轮喷气发动机
Douglas Aircraft Company	(美国)道格拉斯公司
dovetail	楔形榫头
dovetail attachment	燕尾式接合
dow jones industrial average	道琼斯工业平均数
dowel	暗孔
downstream effects	下游的效果
draft contracts exchanged	已交换合同草案
drafting standards manual	标准化手册草案
drag coefficient	阻力系数
drain line	放油管路
drain plug	放油塞
drain tube	放油管
drains tank ejector	油箱排放喷射泵
drawing alteration request	图纸变更要求
drawing die	拉模
drawing introduction sheet	图纸说明表
drawing term	制图术语
Dreamlifter	梦幻运输机(波音 747 - 400LCF)
Dreamliner	梦幻客机(波音 787)
dressed engine	带附件管道和短舱的发动机
dressed spare engine	含短舱等外部设备的备用发动机
dressing	外部设备
drip stick	量油尺

drive arm	盘鼓
driven fan	驱动风扇
driven unit	传动装置
driving improvements	驱动改进
driving sprocket	驱动链轮
driving torque	驱动扭矩
drop angle	倾角
drop dead timing	终止期限
drop load	抛载荷
drop test	跌落试验
drum	鼓筒
dry lean emission	干式低排放
dry lean emission combustor	干式低排放燃烧室
dry low emission	干式低排放
dual alloy discs	双合金轮盘
dual annular combustor	双环腔燃烧室
dual dome combustor	双半球燃烧室
dual energy supply starter	可用双能源的起动机
dual fuel combustion system	双燃料燃烧室系统
dual function starter	双功能起动机
dual material titanium alloy friction welded blisk	（欧盟项目）双材料钛合金摩擦焊叶片盘
dual micro disc	两种材料构成的微型涡轮盘
dual-alloy shaft	两种合金材料制成的轴
duct propeller	涵道螺旋桨
ducted afterburner	涵道加力燃烧室
ducted fan	涵道风扇
ducted fan engine	涵道风扇发动机
ducted fan unmanned aerial vehicle	涵道风扇无人机
ducted propeller	涵道螺旋桨
ductile crack	延性裂纹
ductile fracture	韧性断裂
duel-speed drive device	双速传动装置
Duits-Niederlande Windtunnels	德国荷兰风洞中心

dummy compressor	工艺压气机
dummy hot run	模拟热试车
dummy turbine	工艺涡轮
dump diffuser	突扩扩压器
dumping code	反倾销法案
duple fuel spray nozzle	双油路燃油喷嘴
duplex coatings	双层涂层
duplex reduction gearbox	双联减速器
durable repair patch	长寿命修理补丁
dye penetrant checked	经渗色检验过的
dye penetrant inspection	着色检验
dynamic balance	动平衡
dynamic clash checking detection	动态碰撞检测
dynamic head	动压头
dynamic image	动态图像
dynamic measuring car	动态测量车
dynamic parameter measuring and recording system	动态参数测量记录系统
dynamic pressure calibrator	动态压力校准设备
dynamic pressure measurement	动态压力测量
dynamic reduction	(有限元)动态缩减自由度
dynamic stress measurement	动应力测量
dynamic temperature	动态温度
dynamic temperature measurement	动态温度测量
dynamic test	动态实验
dynamo	发电机
dynamo engine	发电机引擎
dynamometer	测功器

E3E programme	(德国)以经济性、效率和环保为目标的发动机研究项目

Eagle	鹰-战斗机(波音公司,美国)
early fault detection of gear crack	齿轮裂纹故障早期检测
earning assets	盈利资产
earning rate	收益率
ease of repair	易维修
easy crew transition training	机组轻松转换训练
easy machining and forming	易于加工和成形
eccentric running	偏心运转
eccentricity of gravity centre	重心偏心度
Eclipse	日蚀-公务机(日蚀飞机公司,美国)
EcoJet	环保节能喷气机
ecology collector tank	环保型滑油箱
eco-management and audit scheme	(欧洲)生态管理和评估体系
economic cruise	经济巡航
economic cruising rating	经济巡航状态
economic fluctuation	经济波动
economic speed	经济速度
economical cruising speed	经济巡航速度
Ecore	通用核心机计划(GE)
eddy current testing	涡流探伤
edge over	保持对……的优势
edge radii	导角
education training & development	教育培训和发展
e-enabled maintenance	电子化维修
e-enabled power tool	电子化工具
effective perceived noise in decibels	有效感知噪声(分贝)
effective perceived noise level	有效感知噪声级
effective perceived noise level in decibels	有效感知噪声级(分贝)
effective pitch	有效螺距
effective power	有效功率
effective pressure	有效压力
effectiveness	效能
efficiency deficit	效率下降
efficiency-reducing air bleed	降效引气

efficient small scale propulsion	高效小尺寸推进
efflux	喷射气流
efflux angle	流出角
efflux constraint	喷流限制
efflux control	喷射控制
efflux pattern	喷流方式
effusion cooling	多孔冷却
eigenfrequency	固有频率
ejector nozzle	引射喷管
ejector pump	引射泵
elastic constraint	弹性约束
elastic deflection	弹性挠曲
elastic disc	弹性盘
elastic element	弹性元件
elastic hysteresis	弹性延滞
elastic modulus	弹性模量
elastic sleeve bearing	弹性套筒轴承
elasto hydro-dynamic	弹性流体动力学
elastomer tip	弹性尖顶
elastomeric rubber materials	弹性橡胶材料
elasto-plastic disc	弹-塑性盘
Eldim BV	(法国)艾尔迪姆公司
electric actuator	电力制动装置
electric arc welding	电弧焊接
electric breather	电动油气分离器
electric drive system	电力驱动系统
electric explosive valve	电爆阀
electric generator	发电机
electric motor cooler	电机冷却器
electric pump	电动泵
electric spark	电火花
electric starter	电起动机
electrical back-up hydrostatic actuator	液传作动器的电传备份
electrical discharge drilling machine	电火花打孔机

electrical grid	电网
electrical grid system	电网系统
electrical harness interconnects	电缆连接
electrical harnesses	电缆
electrical ice protection	电加温防冰
electrical parameter drift	电参数漂移
electrical power generation and distribution systems	发电和供电系统
electrical starter	电起动机
electrical thrust reverser actuation system	电作动反推系统(赛风)
electrically-driven gyroscope	电动陀螺仪
electrically-heated windscreen	电热挡风玻璃
electrifying progress	电气化进展
electro chemical machining	电化学加工
electro discharge machinetool	电火花加工机床
electro discharge machining	电火花加工
electro hydraulic servo valve	电液压伺服阀
electro hydrostatic actuator	电子液压作动器
electro magnetic compatibility	电磁兼容性
electro magnetic interference	电磁干扰
electro magnetic susceptibility	电磁敏感性
electro mechanical actuator	机电作动器
electro motive force	电动力
electro operating instructions	电动操纵说明书
electro static sensitive devices	静电敏感装置
electro-acoustic actuation	电声传动
electrochemical corrosion	电化学腐蚀
electrochemical corrosion migration	电化学腐蚀迁移
electrochemical machining	电化学加工
electrod hydraulic proportional closed-loop control structure	电液比例闭环控制结构
electro-driving accessories	电气化附件传动
electro-hydraulic actuator	电动液压制动装置

electrohydraulic proportional flow control valve	电液比例调速阀
electro-hydraulic valve	电动液压活门
electrolytic cleaning solutions	电解清理溶液
electromagnetic interference check	电磁干扰检查
electromagnetic pulse	电磁脉冲
electromagnetic riveting	电磁铆接
electromechanical actuator	机电作动器
electromechanical component	机电元件
electromechanical pumps	电动机械泵
electron beam brazing	电子束钎焊
electron beam heat treatment	电子束热处理
electron beam machining	电子束加工
electron beam welder	电子束焊机
electron beam welding	电子束焊
electron beam-physical evaporation deposition	电子束物理气相沉积
electron diffraction pattern	电子衍射图
electron gun	电子枪
electronic centralized aircraft monitor	飞机电子化集中监控器
electronic components	电子元器件
electronic control unit	电子控制装置
electronic counter measures	电子对抗
electronic data management	电子文件管理
electronic data system	电子数据系统
electronic flight information system	飞行仪表显示系统
electronic fuel pumping and metering system	燃油泵送和计量电子系统
electronic general arrangement	电子化总体结构图
electronic induction gauge	电感量仪
electronic landing system	电子着陆系统
electronic protection measures	电子保护措施
electronic support measures	电子支持措施
electronic torque multipliers	电子扭矩放大器
electronic warfare	电子战
electronics intelligence	电子情报

electro-optical infrared sensor	光电红外传感器
electrophoretic deposition	电泳沉积
electrostatic spray assisted vapor deposition	静电喷涂辅助沉积
elevator	升降舵
elevator trim	升降舵配平
elevon	升降副翼
elongation	延伸率
elongation at break	断裂延伸率
embarking or disembarking passengers	旅客上或下飞机
Embraer regional jet	安博威支线客机(安博威公司,巴西)
Embraer SA	巴西航空工业公司(安博威,巴西)
emergency approach	紧急进场
emergency exit hatch	紧急出口窗
emergency go-around	紧急情况下终止着陆爬升
emergency landing	紧急着陆
emergency response management plan	紧急响应管理计划
emergency response team	应急响应小组
emergent technology	新技术
emerging market	新兴市场(特指中国、印度等)
emission certification requirement	排放取证要求
emission check	排放检验
emission constraints	限排放规定
emissions	排放
emissions trading system	温室气体排放交易体系
empennage	尾翼
empirical formula	经验公式
empowered employee	负责任的雇员
empowerment	授权
en route climb	航线飞行上升
en route climb performance	航线上升性能
en route control	航路管制
enabling propulsion materials	高效推进材料
enabling technology	可实现技术

Encore	恩考尔-公务机(塞斯纳飞机公司,美国)
encountered	遭遇的
encountered starting test	偶然性起动试验
end-bend	端部弯曲
end-bended airfoil	端弯叶片
endothermically fueled scramjet engine flight demonstrator	(美国)吸热超燃冲压发动机飞行验证机
end-to-end	终端对终端
endurance test	持久性试车
end-user buy-in	最终用户购买
endwall boundary layer	顶端壁附面层
endwall contouring	端壁造型
endwall loss	端壁损失
endwall vortex	端壁涡流
endwall vorticity	端壁涡量
energy efficient engine	高效发动机
energy storage modulus	储能模量
energy trends and alternate fuels	能源发展趋势与替代燃料
enforcing combustor performance	加力燃烧室性能
Engine 3E (German government/industry funded program)	(德国政府/工业基金项目)经济高效和环保的发动机
engine acceleration	发动机加速性
engine alarm panel	发动机报警仪表盘
Engine Alliance	发动机联盟公司 (GE/普惠合作,美国)
engine analytic redundancy technique	发动机解析余度控制
engine anti-ice valve	发动机防冰活门
engine backlog	发动机订单
engine base maintenance	发动机基地维修
engine build unit	组装好的发动机
engine business unit	发动机业务部
engine calibration facility	发动机标定设施
engine center of gravity	发动机重心

engine centerline	发动机中心轴线
engine certification	发动机取证
engine certification campaign	发动机取证活动
engine characteristics	发动机特性
engine clearance measurement	发动机间隙测量
engine closed-loop control	发动机闭环控制
engine component improvement	发动机部件改善
engine condition monitoring	发动机条件监视
engine configuration	发动机构型
engine control and monitoring unit	发动机控制和监视装置
engine control monitoring	发动机状态监视
engine control system dynamic digital simulation	发动机控制系统动态数字仿真
engine control system physical simulation	发动机控制系统物理仿真
engine covers	发动机机罩
engine cowl anti-ice	发动机整流罩防冰
engine cradle	发动机架
engine crank switch	发动机曲柄开关
engine cross section	发动机截面图
engine cycle characteristics	发动机循环特性
engine data acquisition and reduction system	发动机数据采集处理系统
engine data center	发动机数据中心(罗罗北美)
engine debris analysis	发动机碎屑分析
engine deceleration	发动机减速
engine design memorandum	发动机设计用文件
engine design point	发动机工作点
engine despatch	发动机发货
engine destructive failure	发动机破坏性故障
engine development and certification programme	发动机开发和取证计划
engine development plan	发动机开发计划
engine development programme	发动机开发规划
engine development programme highlight	发动机开发大事记
engine display unit	发动机显示系统
engine drive pinion	发动机驱动小齿轮

engine driven pump	发动机驱动泵
engine durability failure	发动机耐久性故障
engine electronic control	发动机电子控制器
engine electronic control unit	发动机电子控制系统
engine electronic control 1st engine unit	首套发动机电子控制
engine endurance block test	发动机耐久性台架试验
engine endurance test	整机疲劳试车
engine enhanced package	技术升级型发动机
engine fabrications	发动机钣金件
engine failure	发动机故障
engine failure diagnosis	发动机故障诊断
engine failure rate	发动机故障率
engine failure speed	发动机失效速度
engine family	系列发动机
engine fault tolerant control	发动机容错控制
engine feathered	顺桨
engine fire extinguishing system	发动机灭火系统
engine fire shut-off switch	发动机灭火开关
engine fireseal	发动机防火封严框
engine fire-warning light	发动机火警信号灯
engine flame out	发动机熄火
engine flight cycle	发动机飞行循环
engine flight hour	发动机飞行小时
engine fuel pressure	发动机燃料压力
engine fuel system	发动机燃油系统
engine fumes	发动机排出的烟气
engine gas dynamics	发动机气体动力学
engine hand-turn access	发动机手摇把口盖
engine hardware changes	更换发动机硬件
engine health management	发动机健康管理
engine health monitoring	发动机状态监控
engine ideal cycles	发动机理想循环
engine indicate crew alarm system	发动机指示机组报警系统
engine indicating unit	发动机显示器

engine indication and caution advisory system | 发动机指示与警告提示系统
engine in-flight restart | 发动机空中再起动
engine in-flight shutdown | 发动机空中停车
engine inhibiting | 发动机油封
engine inlet | 发动机进气
engine inlet duct rumble | 发动机进气道隆隆声
engine installation diameter nacelle drag | 发动机安装直径短舱阻力
engine integrity | 发动机一体化
engine intelligent test measuring system | 发动机智能测试系统
engine interface unit | 发动机接口元件
engine life cycle cost | 发动机寿命期费用
engine linearquadratic regulator | 发动机线性二次型调节器
engine low cycle fatigue test | 发动机低周疲劳试验
engine main bearing | 发动机主轴承
engine maintenance cost specification | 发动机维护费用规定
engine maintenance manual | 发动机维护手册
engine management programme | 发动机运营管理计划
engine manual | 发动机手册
engine master switch | 发动机总电门
engine mockup | 发动机木制模型
engine model | 发动机模型
engine model development | 发动机模型的发展
engine modelling | 发动机建模
engine monitoring unit | 发动机监控装置
engine mount bolt | 发动机装机螺栓
engine mounted alternator | 安装在发动机上的交流发电机
engine mounting attachment | 发动机安装节
engine multivariable integral control | 发动机多变量综合控制
engine noise measurement | 发动机噪声测量
engine non-ideal cycles | 发动机实际循环
engine oil system | 发动机滑油系统
engine oil warning light | 发动机滑油警告灯
engine open-loop control | 发动机开环控制

engine operability	发动机操作性
engine operation instructions	发动机操纵手册
engine operational failure	发动机功能故障
engine operating limitation	发动机工作限制
engine order book	发动机订单
engine out standard instrument departure	单发失效时标准仪表离场程序
engine overheat detection	发动机过热探测
engine overheat warning light	发动机过热警告灯
engine performance	发动机性能
engine performance parameters	发动机性能参数
engine pod	发动机吊舱
engine power	发动机功率
engine power ratings	发动机功率设置
engine pressure ratio	发动机增压比
engine pressure ratio downtrim logic	发动机压比下调逻辑
engine pressure ratio uptrim logic	发动机压比上调逻辑
engine rating	发动机额定功率
engine rating limitation	发动机额定功率限制值
engine real-time simulation	发动机实时模拟
engine redundancy control	发动机冗余度控制
engine removal rate	发动机拆换率
engine removal rate for exceedance of performance deterioration	性能衰减超过规定换发率
engine research building	发动机研究大楼
engine response to inlet distorted flow	发动机对进气畸变的响应
engine routine	发动机程序
engine running lines	发动机工作线
engine run-up area	发动机试车区
engine section stator	一级导叶
engine sensitivity to inlet distorted flow	发动机对进气畸变的敏感度
engine sensor	发动机传感器
engine shutdown	发动机停车
engine shut-down capacity	发动机停车能力
engine speed indicators	发动机转速指示器

engine stability margin	发动机稳定性余度
engine stall	发动机熄火
engine stand	发动机托架
engine strength test	发动机强度试验
engine strip	发动机分解
engine structural integrity program	发动机结构完整性计划
engine structural test	发动机结构试验
engine subsystems	发动机子系统
engine supervisory control	发动机管理控制
engine surge	发动机喘振
engine surge margin	发动机喘振裕度
engine suspension frame	发动机吊架
engine suspension system	发动机吊挂系统
engine test facility	发动机试车设备
engine test monitoring	发动机试验监控
engine total care agreement	发动机全面维护协议
engine type certificate	发动机型号合格证
engine validation of noise reduction concepts	发动机降噪方案验证
engine vibration endurance test	发动机振动耐久性试验
engine vibration indicator	发动机振动指示表
engine vibration measurement	发动机振动测量
engine vibration mode analysis	发动机振动模态分析
engine vibration monitor unit	发动机振动监测仪
engine vibration monitoring	发动机振动监控
engine with bird damaged fan	装有被鸟击伤风扇的发动机
engine with low parts count	零件少的发动机
engine-cockpit interface unit	发动机与驾驶舱接口装置
engine-driven generator	发动机驱动的发电机
engine-driven hydraulic pump	发动机驱动液压泵
engineering and manufacturing development	技术与制造开发
engineering change	技术变更
engineering change notice	技术变更通知
engineering change package	技术变更包
engineering change release document	技术变更说明文件

engineering change request	技术变更请求
engineering control authority	工程控制主管机构
engineering critical parts plan	重要工程部件计划
engineering design instruction	工程设计指令
engineering design specification	工程设计规范(罗罗)
engineering drawing	工程制图
engineering inspection specification	工程检查规范
engineering material specification	工程材料规范
engineering process specification	技术流程规范
Engineering Review Board	技术审核委员会
engineering source approval required	工程资源审批要求
engineering specification plan	技术要求计划
engineering standards department	技术标准部
engineering standards specification	工程标准规范
engineering technical report	技术支持报告
engineering work package	工程技术工作包
engineering，procurement and construction	设计、采购和施工
engineering thermodynamics	工程热力学
engineers seconded to...	派往……的工程师
engine-mounted electronic controller	发动机载电子控制器
engine-propeller matching	发动机-螺旋桨匹配
EnginSoft	(意大利)工程软件公司
enhanced durability engine	耐久性增强型发动机
enhanced performance engine	性能提升型发动机
enhanced reference system	空间电子坐标系
enhanced solution	改进的解决方案
enterprise resource management	企业资源管理
enterprise system integration	企业系统集成
enterprise resource planning	企业资源计划
enthalpy	焓
entomopter	(美国微型)昆虫飞机
entrepreneurship	创业
entropy	熵
entry into service	投入运营

entry into service readiness	服役准备
envelope	包络
enveloping line	包络线
environmental barrier coating	环保型热障涂层
environmental compatibility	环境兼容性
environmental conditioning system	环境调节系统
environmental control system	环境控制系统
environmental control system ducting	环境控制系统管道
environmental control system precooler	环境控制系统预冷器
environmental legislation	环境法规
environmental management system	环境管理系统
Environmental Protection Agency	环保局
environmental research and aircraft sensor technology	环境研究与飞机传感器技术
environmental test	环境试验
environmentally friendly aero engine	环保型航空发动机(欧洲联盟资助项目)
environmentally friendly engine program	(欧盟)环保发动机项目
environmentally friendly high-speed aircraft	(欧盟)环境友好型高速飞机项目
environmentally friendly inter city aircraft powered by fuel cells	(欧盟)燃料电池动力环保城际飞机项目
environmentally responsible aviation	绿色航空
epoxy resin	环氧树脂
equi-axed crystal structure	等轴晶体结构
equiaxed grains	等轴晶
equilibrant mechanism	均衡机制
equity	资产净值
equity investment	股本投资
equivalence ratio	当量比
equivalent air speed	当量空速
equivalent condition	等效状态
equivalent divergent angle	当量扩张角
equivalent power	当量功率
equivalent ratio	当量比

equivalent still air distance	等效静风航程
Erieye	爱立眼-预警机(安博威公司,巴西)
erosion resistant	抗腐蚀
error band analysis	误差带分析
error detection	误差检测
escalation criteria	升级标准
estimated approach time	预计进场时间
estimated flight time	预计飞行时间
estimated time en route	预计航线飞行时间
estimated time of departure	预计离港时间
etch inspection	蚀刻检查
etching	蚀刻
EULER recuperated engine (5th EU framework program)	欧拉回热式发动机(欧盟第 5 框架项目)
Eulerian-Lagrangian two phase large-eddy simulation	欧拉-拉格朗日两相大涡模拟
Eurofighter Typhoon combat aircraft	欧洲飓风战斗机
Eurojet Turbo GmbH	(在慕尼黑的)欧洲喷气涡轮公司
Eurojet Turbo GmbH	欧洲喷气涡轮发动机公司(德国/英国/西班牙/意大利)
European Union standardisation report	欧盟标准化报告
European Aeronautic Defence and Space Company	欧洲航空防务和航天公司
European aerospace quality group	欧洲航空航天质量组
European aerospace supplier evaluation	欧洲航空航天供应商评估
European Association of Aerospace Industries	欧洲航空航天工业协会
European Aviation Safety Agency	欧洲航空安全局(原欧洲联合航空局)
European Chemicals Agency	欧洲化学品管理局
European Commission	欧盟
European Economic Community	欧洲经济共同体
European Foundation for Quality Management	欧洲质量管理基金会
European Free Trade Association	欧洲自由贸易联盟
European Helicopter Safety Team	欧洲直升机安全执行小组

European Union	欧盟
European Union framework program	欧盟框架科研计划
European Union Trade Commissioner	欧盟贸易专员
Europrop International GmbH	欧洲动力国际公司（负责 TP400M 涡桨发动机）
Europrop International，Munich	在慕尼黑的欧洲动力国际公司
EU's Seventh Framework Program	欧盟第七框架科研计划
evacuated capsule	真空舱
evaluation of indication	示像评判
ever-decreasing timescales for product development	产品开发时间不断缩短
evidence of compliance	合格证明
Ex Works	工厂交货
excess air coefficient	余气系数
excess air combustion	过量空气燃烧
excessive decent rate	下降率过大
exchange rate	敏感性
exciter box	励磁箱
exciting force	激振力
exducer	出口导风轮
exhaust boiler	废气锅炉
exhaust collector	排气收集器
exhaust diffuser	排气扩压器
exhaust emission	尾气排放
exhaust gas temperature	排气温度
exhaust gas temperature indicator	排气温度指示表
exhaust gas temperature probe	排气温度探头
exhaust gas turbine	排气涡轮
exhaust heat exchanger	废气加温器
exhaust impulse	排气冲量
exhaust jet noise	喷气噪声
exhaust manifold	排气总管
exhaust mixer	排气混合器
exhaust muffler	排气消音器

exhaust nozzle	尾喷管
exhaust nozzle characteristics	尾喷管工作特性
exhaust nozzle control	喷口控制
exhaust nozzle exit	尾喷口
exhaust nozzle external flow	尾喷管外流
exhaust nozzle internal flow test	尾喷管内流试验
exhaust nozzle test	尾喷管试验
exhaust system	排气系统
exhaust unit	排气装置
exhaust unit and thrust reverser	排气系统和反推装置
exhaust unit mounting flange	排气装置安装边
exite guide vane	出口导向叶片
exothermic welding	热焊
expanding temperature transmitter	膨胀式温度敏感元件
expansion coefficient	膨胀系数
expansion muffler	扩引式消声器
expansion ratio	膨胀比
expendable engine	短寿命发动机
expendable turbine engine	一次性涡轮发动机
expensive to retrofit	高成本翻新
experimental clean combustor program	实验性低排放燃烧室计划
explanation of drawing terms	制图术语解释
explosion welding	爆炸焊接
explosive forming	爆炸成形
exponential growth	指数式增长
export administration regulations	（美国）出口行政规定
Export Administration Review Board	（美国）出口管理监管委员会
export control classification number	（美国）出口管制分类编码
extended lean enterprise	扩展的精益企业
extended range	延程
extended range fuel system	延程燃料系统
extended range operations	延程运营
extended range twin operations	双发延程飞行
extended shop visit interval	延长的大修期

external characteristics	外部特性
external compression inlet	外压式进气道
external fluid simulation	(叶片)外流模拟
external heat transfer	外部热传递
external heatshielding	外部隔热层
external power receptacle	外部电源插座
external protocols	外部协议
external reflux	外回流
external spline	外花键
external suppliers	外部供应商
externals	(发动机)外部设备
extra efficient	超级效率
extra wide body（A350）	超宽体飞机(空客 A350)
extreme short takeoff and landing	超短距起降
extremely remote	偏远地区
extructed	挤压成型
eyepiece	目镜

Fabrica Argentina de Aviones	阿根廷飞机制造厂
facility layout	工厂布局
factory cost	制造成本
factory engine test	发动机出厂测试
factory test	工厂试车
fail safe	故障安全
failsafe loadpath	故障保险传力路线
fail-safe structure	故障自动防护结构
failure analysis	故障分析
failure detection/isolation/accommodation	故障检测隔离调整
failure diagnosis	故障诊断
failure effect analysis	故障后果分析
failure indicating system	故障指示系统

failure intensity	失效密度
failure investigation	故障调查
failure likehood	失效频度
failure mechanism	失效机理
failure mode	失效模式
failure modes and effects analysis	故障模式和后果分析
failure probability	失效概率
failure reporting, analysis & corrective actions systems	故障报告分析和排除系统
failure simulation	故障模拟
fair and ethical operation	公正并合乎职业道德的运营
Fairchild Dornier	(德国)仙童道尼尔公司
fairing	整流罩
Falcon	猎鹰-公务机(达索公司,法国)
fall-back position	应急情况
fan	风扇
fan air valve	风扇气阀
fan and booster mapping	风扇和增压器特性图
fan assembly	风扇组件
fan blade	风扇叶片
fan blade clapper	扇叶阻尼凸台
fan blade containment test	风扇叶片包容试验
fan blade leading edge and trailing edge	风扇叶片前缘和后缘
fan blade off	风扇叶片分离
fan blade off analysis	风扇叶片分离分析
fan blade off rig test	风扇叶片分离部件试验
fan blade release kinematic	风扇叶片分离后的运动
fan blade-out load	风扇叶片分离载荷
fan bypass performance	风扇外涵道性能
fan case	风扇机匣
fan cowl door	风扇整流罩舱门
fan cowling	风扇整流罩
fan curvic coupling	风扇圆弧篦齿联轴器
fan design	风扇设计

fan disc	风扇盘
fan disc spacer	风扇盘垫片
fan disk insert	风扇盘隼头
fan drive gear system	风扇驱动齿轮系统
fan exhaust noise	风扇排气噪声
fan exit guide vane	风扇出口导向器
fan flutter	风扇颤振
fan gap/chord ratio	风扇间隙与弦长比
fan imbalance	风扇不平衡
fan integrity and containment risk	叶片整体结构和包容性风险
fan intermediate case	风扇中介机匣
fan noise	风扇噪声
fan nozzle throat	外涵喷管喉道
fan operating line control	风扇工作线控制
fan outlet guide vane	风扇外涵整流叶片
fan periphery	风扇边缘
fan pressure ratio	风扇压比
fan Qa	风扇单位气流量
fan retention shaft	风扇定位轴
fan root efficiency	风扇叶根效率
fan root exit	风扇叶根排气
fan root inlet	风扇叶根进气
fan rotor	风扇转子
fan rotor blade	风扇转子叶片
fan sense debris monitoring	异物撞击风扇碎屑检测（罗罗）
fan speed detector	风扇速度探测器
fan speed probe	风扇速度探头
fan stub shaft assembly	风扇支承轴组件
fan tip efficiency	风扇叶尖效率
fan tip exit	风扇叶尖排气
fan tip inlet	风扇叶尖进气
fan track seal	风扇机匣内衬封严
fan trim balancing	风扇刚性动平衡
fan/intake compatibility	风扇和进气道的兼容性

fan-on-blade	叶片上的风扇
far field noise	远场噪声
fast ferry	高速摆渡船
fast fourier transform	快速傅里叶变换
fast make cell	高速生产单元
fast turnaround	交付周期短
fastener	紧固件
fastener loss	紧固件丢失
fastenerless	无紧固件
fasteners correctly fitted	正确安装紧固器
fastest selling airliner	销售最快的客机
fastest-made	最快的产量
fast-growing market	高增长市场
fast-paced engine test program	发动机的快速试验计划
fat prices	高昂的代价
fatal accidents	致命事故
fatigue crack	疲劳裂纹
fatigue crack growth rate	疲劳裂纹扩展率
fatigue fracture	疲劳破裂
fatigue initiation	疲劳源
fatigue life evaluation	疲劳寿命评定
fatigue load spectrum	疲劳载荷谱
fatigue strength	疲劳强度
fatigue striation	疲劳条带
fatigue test	疲劳试验
fatigue wear	疲劳磨损
fault adjustment tactics	清除故障策略
fault and safety analysis	故障和安全性分析
fault clearance	故障清除
fault code analysis	故障编码分析
fault detection filter	故障检测滤波器
fault detection subsystem	故碍探测子系统
fault diagnosis	故障诊断
fault diagnosis expert system	故障诊断专家系统

fault indication and correction action	故障指示和排除
fault isolation	故障隔离
fault isolation manual	故障隔离手册
fault not found	未发现故障
fault threshold value	故障阈值
fault tree analysis	故障树分析法
faulty part report	报废零件报告
favorable balance of trade	贸易顺差
feathering control	顺桨操纵
feathering pitch	顺桨桨距
feathering pump	顺桨泵
Federal Aviation Administration	美国联邦航空局
Federal Aviation Regulations	美国联邦航空条例
Federal Aviation Regulations 33	(发动机审定条例)美国联邦航空条例 33 部
Federal Aviation Regulations part 36 stage 3 rule	美国联邦航空条例第 36 部第 3 等级标准
Federal Bureau of Investigation	美国联邦调查局
Federal Information Processing Standards (USA)	(美国)联邦信息处理标准
FedEx	联邦快运
feed water heater repowering	用供水型加热器增加功率
feedback gearbox	反馈齿轮箱
feed-back unit	反馈装置
feet per second	每秒英尺
felt reusable surface insulation	(美国国家航空航天局)柔性可重复使用表面绝热材料
ferritic-austenitic microstructure	铁素体-奥氏体微观结构
ferrograph oil analysis	滑油铁谱分析
ferrous powder metallurgic material	粉末冶金铁基材料
FE-SAFE	(ANSYS 公司的)结构疲劳分析软件
fewer disruptions	中断少
Fiat Avion	(意大利)菲亚特-艾维欧公司
FiatAvio	(意大利)菲亚特航空工程公司

fiber metal	纤维金属
fiber optic sensor	光纤传感器
fiber reinforced plastics	纤维加强塑料
fiber stress	纤维应力
fiber-braided composite fan case	纤维编织复合材料风扇机匣
fiber-metal laminates	纤维金属层压板
fiber-metal sandwich panel	纤维金属夹层板
fiberscope	光纤内窥镜
fibre insulation	纤维隔层
fibre placement	纤维布置
fibre reinforced rings	纤维增强环
fibre surface treatment	纤维表面处理
fibrous insulating material	纤维隔热材料
fibrous refractory composite insulation	（美国国家航空航天局）碳纤维复合绝热材料
field failure	现场失效
field maintenance	现场维修
field mission	现场任务
field rep support	驻场代表支援
fighter	战斗机
fighter engine team	战斗机发动机工作组
Fighting Falcon	战隼-战斗机（通用动力/洛克希德公司,美国）
filler cap adapter	加油口盖接口
filler metal	熔填金属
filler temperature transmitter	充填式温度传感器
fillet radii	倒角半径
film cooling	气膜冷却
film cooling blade	气膜冷却叶片
film cooling holes	气膜冷却孔
film effectiveness	气膜效率
filter by-pass valve	油滤旁通活门
filter drain valve	油滤放油活门
filtration	过滤

final approach speed	最后进场速度
final assembly hall/shop	总装车间
final assembly line	总装线
final rupture	最终断裂
finance	金融
financial incentire	财政激励
financial statement	财务报表
financing corporation	金融公司
fine grain bulk artificial graphite	人造细晶粒整体石墨
finger lift	指提钮
finished goods inventory	成品存货
finished part stores	成品零件库
finished ring mount	成品环形联结件
finite cascade	有限叶栅
finite element analysis	有限元分析
finite element method	有限元法
fir tree root	枞树形叶根
fire containment	防火包容
fire containment ring	防火包容环
fire detection	火警探测
fire extinguisher bottle	灭火器瓶
fire extinguishing	灭火
fire precaution	防火
fire test	着火试验
fireproof bulkhead	防火隔板
firing	焙烧
firing contact pin	点火撞针
firm engine order	发动机固定订单
firm-fixed price contract	定价合同
first article inspection	首件产品检验
first article inspection report	首件产品检验报告
first core engine run	首次核心机运转
first engine test team	首次发动机试车工作组
first flight	首飞

first full engine run	首次整机运转
first full-engine test	首次整机试车
first in first out	先进先出
first run	首次运行
first stage retaining toothed ring	第一级固定齿圈
first-name teams	以小名互称的小组
first-pass-yield	试车一次性通过率
Fischer-Tropsch	(生产合成代用燃料的)费舍尔-特劳什工艺
Fischer-Tropsch fuel	(由碳氢化合物转化而成)弗舍尔-特劳什燃油
fixed assets	固定资产
fixed assets accounting	固定资产核算
fixed cascade	固定叶栅
fixed cost engine maintenance	发动机维修成本固定
fixed cowl	固定舱门
fixed pitch propeller	固定螺距螺旋桨
fixed pitch stop	固定螺距止动销
fixed process document	生产流程管控文件
fixed wing	固定翼
fixed wing aircraft	固定翼飞机
fixed-base operation	固定运营基地
flame flash-back	(指火焰经传导后)火舌回闪
flame front	火焰前锋
flame holder	火焰稳定器
flame holder test	火焰稳定器试验
flame jet gradient test	火焰喷射梯度试验
flame out	(发动机)熄火
flame propagation	火焰传播
flame stabilizer	火焰稳定装置
flame tube	火焰筒
flame tube cooling	火焰筒冷却
flammability	可燃性
flammable fluid	可燃液体

flange	法兰盘
flange load	法兰载荷
flap	襟翼
flap blowing engine	襟翼吹气发动机
flap carriage	襟翼托架
flap extension	放襟翼
flap position transducer	襟翼位置传感器
flap retraction	收襟翼
flap torque tube	襟翼扭力管
flap track	襟翼导轨
flare	飞机着陆前拉平（飞行）
flare	燃烧室扩张段
flared divergent section	喇叭形扩散段
flash butt welded ring	闪光对焊环
flashboard valve	插板阀
flashing beacon	闪光灯
flat profit	低增长利润
flat rated/rating	最大允许推力设置
flat-headed screw	平头螺钉
flat-plate boundary layer	平板边界层
flattening test	压扁试验
flawless delivery	无缺陷的交付
flawless engine performance	完美的发动机性能
flawless new product introduction	无缺陷的新产品引入
flawless quality	无缺陷的质量
fleet capacity	机队总运力
fleet communality	机队通用性
fleet hour agreements	按机队飞行小时付费协议
fleet stagger	机队梯次使用
flexible hose	软管
flexible manufacturing system	柔性制造系统
flexible rotor	柔性转子
flexible support	弹性支承
flexible technical endoscopes	柔性工程内视镜

flexible wall	柔壁
flexing shed ice	挠曲除冰
flexural strength	弯曲强度
flight accuracy	飞行航线准确度
flight altitude	飞行高度
flight cert prop system	推进系统飞行证书
flight certification modification	飞行审定变更
flight compartment	驾驶舱
flight constraint analysis	飞行限制分析
flight control computer	飞控计算机
flight crew cost	飞行机组成本
flight crew operating manual	机组人员操作手册
flight cycle	飞行循环
flight data display	航班数据显示
flight data processing system	飞行数据处理系统
flight data recorder (black box)	飞行数据记录器（黑匣子）
flight deck	飞机驾驶舱
flight deck indicators	驾驶舱指示器
flight envelope	飞行包络线
flight flare	飞行照明弹
flight hour agreement	保修协议（GE）
flight hours	空中飞行小时
flight idle	空中慢车
flight idle speed	飞行慢车速度
flight length	飞行时间
flight level	（标准气压）高度层
flight loads testing using pressure belts	使用压力带的飞行载荷试验
flight Mach number	飞行马赫数
flight management system	飞行控制系统
flight manual	飞行手册
flight meteorological service	飞行气象服务
flight mission analysis	飞行任务分析
flight mode	飞行状态
flight mode annunciator	飞行状态指示器

flight modification approval	飞行变更批准
flight operational quality assurance	飞行运营质量保证
flight operations telex	飞行运营通报
flight ops training/guidance	飞机操作培训和指导
flight path angle	飞行轨迹仰角
flight propulsion system	飞行推进系统
flight readiness review	待飞审查
flight service station	飞行服务站
Flight Standardization Board	飞行标准化委员会
Flight Standards and Airworthiness Division（Civil Aviation Department）	飞行标准及适航部（民航处）
flight test engine	试飞用发动机
flight test programme	试飞项目
flight time	空中飞行时间
flight-development aircraft	试飞用飞机
flight-development testing	试飞试验
flight-line maintenance	航线维护
flight-tested	经飞行试验的
flinger ring	甩油环
float fuel level gauging meter	浮子杠杆式燃油油量表
floating production storage unit	浮动生产储油船
floating production unit	浮动采油船
floating production，storage and offloading vessel	浮式石油生产储存装卸船
floating ring seal	浮环密封
floatwall combustor	浮壁燃烧室
floor space	厂房面积
floor-fixture stool	地面托架
flow adjust/close equipment	流量控制与关断装置
flow blockage	流量堵塞
flow calibrator	流量校验台
flow capacity loss	流量损失
flow distortion	流场畸变
flow divider valve	分流阀

flow modelling	流动模拟
flow of processes	工艺过程
flow path	流程
flow penetration depth	气流穿透深度
flow probe	流量探针
flow production concept	流水线生产方案
flow rate measurement	流量计量测试
flow reversal	回流
flow separation	气流分离
flow sheet	流程图
flow variation	气流变化
flow visualization	流动显形
flowchart	流程图
flowline operating principle	流线型运营原则
Fluent	计算流体动力学软件
fluid heat exchange	流体热交换
fluid hoses	液体管道
fluid schematic	流路图
fluid-cooled scramjet	液体冷却超燃冲压发动机
fluidics amplifier	射流式放大器
fluid-structure interactions	流体与结构相互作用
fluorescence analysis	荧光分析
fluorescent ink	荧光涂料
fluorescent penetrant inspection	荧光渗透检测
fluorescent penetrant inspection line	荧光渗透检测线
fluorescent test	荧光测试
flush bolt	平头螺栓
flutter	颤振
flutter damper	颤振阻尼器
flutter margin	颤振裕度
flutter mode control	颤振模态控制
flyaway cost	整机成本
fly-by-light system	光传操纵系统
fly-by-wire system	电传操纵系统

flying mammoth	飞行毛象(空客 A320)
flying route	飞行航线
flying test bed	飞行试车台
flying trial	飞行试验
flying wing	飞翼
flyover-cutback	(为降噪)减油门起飞
focused factory	专业化工厂
focused ion beam	聚焦电子束
Fokker	(荷兰)福克飞机公司
Fokker Elmo	(荷兰)福克-埃尔莫公司
follow-on test and evaluation	继续测试和评估
force field analysis	受力区分析
forced landing	迫降
forced lubrication	压力润滑
forced lubricator	压力润滑器
forced mixer	强制混合器
forced mixer with common exhaust nozzle	强制混合器与排气喷管为一体
forced response	强迫振动响应
forecooler	预冷气
Foreign Corrupt Practices Act	(美国)海外反腐败法
foreign currency balance	外汇平衡
foreign exchange loss	外汇兑换损失
foreign object damage	异物损伤
foreign object damage	外来物损伤
foreign object debris	异物碎屑
foreign object ingestion	异物吸入
foreign object ingestion test	异物吸入试验
foreign object strike	异物撞击
Foreign Trade Organization	国际贸易组织
foreign-funded enterprise income tax	外商投资企业所得税
forged aluminium	锻造铝制件
forged disc	锻造的盘
forged Jethete (steel) casing	材料为杰赫不锈钢的锻造机匣
forged outer case	锻造外部机匣

forged steel	锻钢
forged surface	锻造表面
forging	锻件
forging die	锻模
forging hammer	锻锤
forging method of manufacture data sheets	生产数据表中的锻造法
formal demonstration approach	正式的验证方法
formula ingredients	配方成分
Fortin barometer	福丁气压表
forward bulkhead	前舱壁
forward mount	前安装节
forward mounting bracket	前安装节托架
forward planning programme	远期计划
forward sweep	前掠
forward swept fan	前掠风扇
forward thrust	向前推力
forward thrust load	前推力载荷
forward thrust reverser controls	前反推装置操纵
forward-looking company	具远见的公司
forward-sweeping wing	前掠机翼
fossil fuel	化石燃料
foundation bolt	地脚螺栓
foundry and machine shop	铸造和机加工车间
foundry defect	铸疵、铸造缺陷
foundry sand	铸砂、铸造用砂
four nozzle deflector	四喷管偏转器
four-blade propeller	四叶螺旋桨
Fourier decomposition	傅里叶分解
Fourier transform infrared spectroscopy	傅里叶红外光谱
four-stroke engine	四冲程发动机
four-stroke piston engine	四冲程活塞发动机
Foxjet	福克斯喷气机-公务机（福克斯喷气机公司，美国）
fracture critical component	断裂临界部件

fracture stress	断裂应力
fracture yield	断裂屈服
framework program	(欧盟科研)框架项目
Frank Whittle	(英国喷气发动机发明者)弗朗克·魏特勒爵士
Fredrickson	(波音所属)弗雷德里克松公司
free alongside ship	船边交货
free carrier	承运人交货
free jet simulated altitude test	自由射流模拟高空试验
free of charge	免费
free on board	离岸价格
free power turbine	自由动力涡轮
free stream property	自由流特性
free stream tube area	自由流管面积
free trade area	自由贸易区
free turbine	自由涡轮
free ventilation	自由通风
free vortex	自由涡
free windmilling	自由风车状态
freeze	冻结
freight tonne kilometers	货运吨公里
freighters	货机
French Defense Procurement Agency	法国国防采办局
French-built engines	法国生产的发动机
frequency decomposition	频率分解
frequency domain analysis of dynamic data	动态数据频域分析
frequency response	频响
frequency spectrum analysis	频谱分析
frequency to direct current convertor	频率直流变换器
frequent and extensive reconditioning	繁多的调试
frequent malfunction emergency analysis	常见故障紧急分析
frequently asked questions	常提出的问题
fretting	微磨损
fretting corrosion	磨损侵蚀

fretting fatigue	磨损疲劳
fretting wear	微动磨损
friction bonding	摩擦联结
friction factor	摩擦系数
friction form resistance	摩擦阻力
friction locked adjuster	摩擦锁定调节器
friction presse	摩擦压力机
friction stir welding	摩擦搅拌焊
front bearing chamber	前轴承腔
front desk team	前台工作组
front fan engine	前风扇发动机
front fan turbo-jet	前风扇涡轮—喷气发动机
front mount	前安装节
front mount pylon interface	前安装节结合部
front split casing	前部整流机匣
front stub shaft	前短轴
fuel & control system bench test	燃油和调节系统整机试验
fuel air ratio	油气比
fuel atomisation	燃油雾化
fuel atomization fineness	燃油雾化细度
fuel atomization uniformity	燃油雾化均匀度
fuel atomizer	燃油雾化器
fuel atomizer injector	燃油雾化喷嘴
fuel boost pump	燃油增压泵
fuel booster pump	升压燃油泵
fuel burn	耗油量
fuel burner	燃油燃烧器
fuel cavitation	燃油中的气穴
fuel cell based auxiliary power unit	燃料电池辅助动力系统
fuel cell stack	燃料电池堆
fuel coefficient	燃油系数
fuel coking	供油开关
fuel concentration	燃油浓度
fuel concentration distribution	燃油浓度分布

fuel consumption	燃油消耗
fuel consumption retention guarantee	耗油率衰减保证
fuel contamination	燃油污染
fuel control unit	燃油调节器
fuel cooled oil cooler	用燃油冷却的滑油冷却器
fuel density	燃油密度
fuel density compensation	燃油密度补偿器
fuel depot	燃料库
fuel distribution system	燃油分配系统
fuel distributor	燃油分布器
fuel droplet size distribution	油滴尺寸分布
fuel dumping	应急放油
fuel duty	燃油税
fuel efficient airplane	节省燃料的飞机
fuel endurance	续航时间
fuel farm	油库
fuel feed system	燃油供给系统
fuel filter	燃油过滤器
fuel filter differential pressure sensor	燃油过滤器压差传感器
fuel flow calibrator	燃油流量校验台
fuel flow distributor	燃油流量分配器
fuel flow divider valve	燃油流量分配器活门
fuel flow gauge	燃油流量表
fuel flow governor	燃油油量调节器
fuel flow measurement	燃油流量测量
fuel flow regulator	燃油流量调节器
fuel flowmeter	燃油流量表
fuel gas desulphurization	燃气脱硫
fuel gauge	燃油油量表
fuel heat value	燃料热值
fuel heating	燃料加热
fuel injection ring	喷油环
fuel injector	喷油器
fuel leakage	燃油渗漏

fuel level annunciator	燃油油量信号器
fuel load	载油量
fuel management system	燃油管理系统
fuel manifold	燃油总管
fuel mass flow	燃料流量
fuel metering orifice	燃油表气口
fuel metering unit	燃油表
fuel metering valve	燃油计量阀门
fuel nozzle test	燃油喷嘴试验
fuel oil heat exchanger	燃油滑油热交换器
fuel pre-ignition	燃油预燃
fuel preprocessing	燃料预处理
fuel pressure pump	燃油增压泵
fuel pressure warning light	燃油压力下降警告灯
fuel property	燃油性质
fuel pump	燃油泵
fuel pump combined with fuel metering unit	一体化的燃油泵和燃油表
fuel quantity indicating compensator probe	燃油量指示补偿器探头
fuel quantity indicating densitometer	燃油量指示密度计
fuel quantity indicating probe	燃油量指示传感器
fuel quantity indicator	燃油量指示器
fuel receptacle	加油接头
fuel recirculation	燃料再循环
fuel reserve	燃油储备
fuel return valve	燃油回油阀
fuel sample	燃料样品
fuel shutoff valve	燃料断流活门
fuel spiking	瞬间过量供油喘振
fuel spray nozzle	燃油喷嘴
fuel swirler	燃油涡流器
fuel tank	燃油箱
fuel tank arrangement	燃油箱分布
fuel tank bladder	燃油箱挤压囊
fuel tank filler cap assemblies	燃油箱加油口盖

fuel tank inerting system	燃油箱惰化系统
fuel temperature gauge	燃油温度表
fuel temperature probe	燃油温度传感器
fuel totalizer	总燃油量表
fuel trimming device	燃油微调装置
fuel volume constraints	燃料量限制
fuel/air ratio	油气比
fuel/oil cooler	以燃油为冷媒的滑油冷却器
fuel/oil heat exchanger	用燃油冷却滑油的热交换器
fuel/oil heat exchanger bypass valve	燃油-滑油热交换器涵道阀门
fuel-air combustion starter	煤油空气燃烧式起动机
fuel-cell power supply	燃料电池动力
fuel-rich mixture	富油混合器
fulcrum vibration	支点振动
fulfillment	执行
full annular rig test	全环型部件试验
full authority digital electronic control system	全权限数字电子控制系统
full authority digital engine control	全权限数字发动机控制
full authority digital engine control system leadtime	发动机全权限数字发动机控制系统交付周期
full core fairing	全核心机整流罩
full cowl nacelle	混合流短舱
full expansion	完全膨胀
full functions	功能齐全
full life test	全寿命试车
full load	满载
full operational capability	具有全面运营能力
full power	最大推力
full power take off	最大油门起飞
full power trial	满功率试车
full power without afterburner	无加力最大推力
full rate production	满负荷生产
full reverse	全油门反推
full scale fan case	全尺寸风扇机匣

full speed	全速
full supply chain capability	完整的供应链能力
full vacuum electron beam welder	高真空电子束焊机
full-feathering propeller	全顺位螺旋桨
full-flap configuration	襟翼全放状态
full-mould casting process	实型铸造
full-power	满功率
full-scale annular combustor test	全尺寸环形燃烧室试验
full-scale engine research	全尺寸发动机研究
full-scale turbine test	全尺寸涡轮试验
full-speed	全速
fully automated data acquisition system	全自动数据采集系统
fully cowled core	全封闭式核心机
fully faired bypass duct	全整流外涵道
fully fledged low pressure turbine risk and revenue sharing partnership	杰出的低压涡轮风险收益共担伙伴关系
fully installed engine	完整装机的发动机
fully integrated digital workflow	全集成数字工作流程
fully vented bearing chamber	全通气式轴承腔
full-year earnings	全年收入
full-year profits	全年利润
fully-integrated powerplant system	一体化推进系统
fume extraction	抽烟
function of flow vs altitude for oil pump	滑油泵高空特性
function of flow vs rotating speed for oil pump	滑油泵转速流量特性
functional interface control document	界面控制功能文件
functional test	功能试验
functional test bed	功能试车台
funding and liquidity	资本融资与流动性
fuse and gap	轴承支架保险栓及其间隙
fuse bolts/flange	保险阀/法兰
fuse design	轴承支架保险栓的设计
fused deposition modeling	熔融沉积成型工艺

fuselage boundary layer diverter	机身附面层分离器
fuselage composite	(法国政府资助的)全复合材料机身
fuselage engine	安装在机身上的发动机
fuselage frame	机身骨架
fuselage intake	机身进气道
fuselage mounted	机身安装(发动机)
fuselage skin	机身蒙皮
fusible pattern extractive cast	熔模精铸
fusion welding	熔焊
fussy logic	模糊逻辑
future affordable turbine engine	(美国)未来经济可承受涡轮发动机项目
future carrier-borne aircraft	(英国)未来舰载机
future joint combat aircraft（JSF）	(美国)未来联合战斗机
future offensive air system	(英国)未来空中攻击系统
future propulsion system	未来推进系统
future strategic tanker aircraft	(英国)未来战略加油机
future technology trend	未来的技术趋势
fuzzy analytic hierarchy process	模糊层次分析法
fuzzy clustering analysis	模糊聚类方法
fuzzy logic technology	模糊逻辑技术

Galaxy	银河-公务机(湾流宇航公司,美国)
Galeb	海鸥-攻击机(索科公司,南斯拉夫)
galling	咬合
gallons per hour	每小时加仑
gallons per minute	每分钟加仑
Gamma titanium-aluminium alloy	伽马钛-铝合金
gap effects	间隙影响
gas analysis	燃气分析
gas constant	气体常数

gas discharge	燃气排气管
gas discharge device	气体放电器件
gas dispenser	加气机
gas dynamics	气体动力学
gas filled detector	充气探测器
gas flow path	气流通道
gas flow pattern	燃气流的流型
gas fuel	气体燃料
gas gathering	天然气采集
gas generator	燃气发生器
gas metal arc welding	气体保护金属级电弧焊
gas property	气体性质
gas separator	气体分离器
gas turbine engine	燃气涡轮发动机
gas turbine generator set	燃气涡轮发电机组
gas turbine manufacturing skill	燃气涡轮制造技术和经验
gas turbine starter	燃气涡轮起动机
gaseous helium	气态氢
gaseous oxygen	气态氧
gas-insulated	气体绝缘
gasoline engine	汽油引擎
gas-shielded arc welding	气体保护焊
gate readers	登机口读卡机
gateway review	入门审查
gating	选通
gating process	选通流程
gauge calibrating status	测量仪器的校准状态
gauge calibration	测量标定
gauge capability	计量能力
gauge pressure	表压
gauze filter	滤网油滤
GE Aircraft Engines	通用电气集团飞机发动机公司
GE Capital Aviation Service	通用电气集团金融航空服务公司
gear alloy	齿轮钢

gear carrying capacity	齿轮承载能力
gear cooling system	齿轮冷却系统
gear oil	齿轮油
gear pin	齿轮销、传动销轴
gear pump	齿轮泵
gear ratio	齿轮传动比
gear train drive	齿轮链系传动
gearbox	齿轮箱
gearbox altitude test	减速器高空试验
gearbox back to back test	减速器直接对比试验
gear-driven dynamics	齿轮传动动力学
geared engine	齿轮减速风扇发动机
geared fan	齿轮驱动风扇
geared oil pump	齿轮油泵
geared turbine engine	齿轮减速涡轮发动机
geared turbofan	齿轮减速涡扇发动机
Gem	格姆发动机(罗罗公司,英国)
General Administration of Civil Aviation of China	中国民航总局(现名)
general agreement on tariffs and trade	关税与贸易总协定
general arrangement	(发动机)总体结构图
general arrangement of aero-engine	发动机总体布局
general arrangement parts list	总体结构零件目录
general aviation	通用航空
General Aviation Manufacturers Association	通用航空生产商联盟
general aviation propulsion	通用航空推进系统
General Electric	通用电气公司(GE)
General Electric next-generation	GE为波音787研制的发动机
general emergency alarm system	一般应急告警系统
General Equipment Department of Chinese People's Liberation Army	中国人民解放军(原)总装备部
general management	综合管理
general performance access	总体绩效评估
general performance-control capability	总体绩效控制能力

general specification	一般规定
general specification for military engines	军用发动机通用规范
general-purpose aircraft	通用飞机
generic certification requirement	通用合格审定要求
generic engine	通用发动机
generic skills	通用技术
generic study model	通用研究模型
genset	发电机组
geometric height	几何高度
geometrical parameters of cross section of aerofoils	叶型截面几何参数
geometrical tolerance	几何公差
geo-potential altitude	地势高度
German Aerospace Center	德国航空宇航研究中心
getting rid of obstacle	排除障碍
Ghost	幽灵发动机(罗罗公司,英国)
glare	(荷兰生产的用于空客 A380 机身蒙皮的玻璃纤维夹层板)格拉尔
glass cloth layers	玻璃布层
glass fibre	玻璃纤维
glass fibre cone	玻璃纤维锥头
glass reinforced laminate（glare）	玻璃纤维增强叠层板
GLAss-REinforced fiber metal laminate	格拉尔叠层版(荷兰代尔夫特理工大学开发的玻璃纤维金属混合夹层板)
glass-reinforced polyester laminate	玻璃纤维增强的聚酯塑料层压板
glide ratio	滑翔比
Global Aerospace Centre for Icing and Environmental Research	全球航空结冰与环境研究中心(罗罗)
global component repair services	全球部件维修服务
global conference for manufacturing	全球制造大会
global coordinating enviroment	全球协作环境
global customer services network	全球客户服务网络
global domination	主导世界

Global Employee Council　　　　　全球雇员委员会

global engineering　　　　　　　　全球工程

global engineering method　　　　全球统一技术方法

global engineering reference　　　全球技术参考资料

global experimental process　　　全球化实验流程（罗罗）

Global Express　　　　　　　　　　环球快车–公务机（庞巴迪公司，加
　　　　　　　　　　　　　　　　　　拿大）

Global Flyer　　　　　　　　　　　环球飞行者号飞机

Global Hawk　　　　　　　　　　　全球鹰–无人机（诺斯洛普–格鲁门公
　　　　　　　　　　　　　　　　　　司，美国）

global material solutions　　　　（普惠公司专为 CFM56 生产备件的）
　　　　　　　　　　　　　　　　　　全球化备件方案

Global Observer　　　　　　　　（液氮燃料电池无人机）全球观察者

global positioning system　　　　全球定位系统

Global Process Quality System Board　全球质量系统流程委员会

Global Procurement Council　　　全球采购理事会

global product development　　　全球产品研发模式

global purchasing　　　　　　　　全球采购

global scorecard　　　　　　　　　全球记分卡

global strategy　　　　　　　　　　全球战略

global supplier portal　　　　　　全球供应商门户

global supply chain　　　　　　　全球供应链

global trade management　　　　国际贸易管理

global vibration of engine　　　　发动机整机振动

global warming　　　　　　　　　全球变暖

globallink data network　　　　（美国 ARICN 公司的）全球数据互
　　　　　　　　　　　　　　　　　　联网

Globemaster　　　　　　　　　　环球霸王–运输机（麦道公司，美国）

glo-crack　　　　　　　　　　　　荧光示裂法

glow discharge mass spectrometry　辉光放电质谱

glyph chart　　　　　　　　　　　雷达图（多轴平面图）

Gnome　　　　　　　　　　　　　诺姆发动机（罗罗公司，英国）

go around　　　　　　　　　　　　复飞

go around thrust　　　　　　　　复飞推力

Gold Care	(波音公司售后服务)金色维护
Golden Eagle	金鹰-战斗机(韩国航太工业集团, 韩国)
good handling and reliability	良好的操作性和可靠性
Goodman diagram	古德曼疲劳图
Goodman's law	古德曼定律(疲劳特性)
Goodrich Aerospace	古德里奇航宇公司(联合技术公司, 美国)
goods receiving/received	收货
goods receiving input process	收货登记流程
goodwill	商誉
Goshawk	苍鹰-教练机(波音公司,美国)
gouge	凿伤
Government Accountability Office	(美国)政府审计署
government engine business	政府发动机业务
government funded equipment	政府资助设备
government loans	政府贷款
government owned contractor operated	国有民营的管理模式
government subsidies	政府财务补贴
governor pressure	调节器压力
grade of tolerance	公差等级
grain boundary	晶界
grain misalignment	晶粒错位
grain orientation diagram	晶粒取向图
grain size	晶粒大小
grain structure	晶粒结构
graphic programming	可视化编程
graphical user interface	图形用户界面
graphite blanket	石墨毡
graphite crucible	石墨坩埚
graphite electrode	石墨电极
graphite sleeve	石墨套
green engine demonstrator	环保型发动机验证机
green investment scheme	绿色投资计划

green manufacturing	绿色制造
greenhouse gas	温室气体
grid-aided engineering	网格辅助工程
grinding	研磨
Gripen	鹰狮-战斗机(萨博公司,瑞典)
gross assets	总资产
gross calorific value	总热值
gross domestic product	国内生产总值
gross heating value	总热值
gross liabilities	总负债
gross margin	毛利
gross national product	国民生产总值
gross profit	总利润
gross profit from sales	销售总利润
gross take-off weight	起飞总重量
gross thrust	总推力
gross weight	总重
ground de-icing	地面除冰
ground development test	地面研发试验
ground effect	地面效应
ground effect vehicle	地面效应车辆
ground effect wing	气垫机翼
ground equipment	地面设备
ground handling	地面操作
ground handling agent	地勤服务代理人
ground handling service	地勤服务
ground idle	地面慢车
ground power control unit	地面电源控制器
ground power unit	地面电源
ground rig testing	地面部件试验
ground run	地面开机
ground run-up enclosure	地面试车准备区
ground service plug	地面电源插头
ground speed	对地飞行速度

ground start	地面起动
ground start cart	地面起动机车
ground strap	接地母线
ground suction lift	地面衬层升力
ground support equipment	地勤支持设备
ground support system	地面支援系统
ground test	地面测试
ground testing	地面测试
ground time	飞机在地面时间
ground training	地面训练
ground transport	地面运输
ground-based air defences	陆基防空系统
ground-based flight simulator	地面飞行模拟器
ground-breaking	奠基
ground-breaking technology	突破性技术
grounded	停飞的
group quality procedure	团队质量程序
group technology	成组技术
growing tendency analysis	(推力)增长趋势分析
Growler	咆哮者-电子战机(波音公司,美国)
growth capability	(推力)增长能力
growth potential	增长潜力
GT Garrett	(美国)盖瑞特公司
guaranteed availability	保障供应
Guest Keen & Nettlefolds Aerospace	吉凯恩(客人-敏锐-内特尔福德)航宇公司
Guggenheim Aviation Partners	(美国飞机租赁公司)古根海姆航空伙伴
guide vane	导流片
guided plug-in cover	先导插装盖板
Gulfstream Aerospace Corporation (Gulfstream)	(美国)湾流宇航公司
Gulfstream GV	湾流 V 型公务机
Gulfstream	湾流-公务机(湾流宇航公司,美国)
gully depth	翼下间隙

gust	阵风
gust force	阵风力
gust load	阵风载荷
gyro horizon	陀螺地平仪
gyrocompass with coupled autopilot	附有自动领航器的电罗经
gyrodyne concept	旋翼飞机
gyroplane	自转旋翼机
gyroscope	陀螺仪
gyroscopic load	哥矢力（或陀螺载荷）
gyroscopic moment	陀螺力矩
gyroscopic procession	陀螺进动

hailstone ingestion test	吸冰雹试验
Hamilton Sundstrand Company	（美国）汉胜公司
hammer riveting	钉锤铆接法
hand bumping	手工锤击
hand forming	手工成形
handing quality	操作品质
handling bleed closed	放气阀关闭
handling bleed valve	控制放气阀
handling bleed valve solenoid block	放气控制阀螺线管箱
handling bleeds	操纵放气
handling tool	装卸工具
hang	悬挂
hanging system	悬挂系统
Hans von Ohain	（德国喷气发动机发明者）汉斯·冯 欧海茵博士
hard conversion	硬性换算
hard fault of control system	控制系统硬件故障
hard particles soldered to the tips	硬颗粒被焊接到叶尖上
hardened layer	硬化层

hardening	硬化
hardening furnace	淬火炉
Hardness-Brinell	（瑞典人 Brinell 定义的）布里涅尔硬度
Hardness-Vickers	（Vickers 定义的）威氏硬度
hardness test	硬度试验
hardware implementing redundancy control	硬件实现冗余度控制
harmonic analysis	谐波分析
harness routing	布线
Harrier	鹞式-垂直起降战斗机（英宇航公司，英国）
harsh field conditions	在外场恶劣环境下
Hawk	鹰-战斗机（英宇航公司，英国）
Hawker	豪客-公务机（豪客比奇飞机公司，美国）
Hawker Siddeley	豪客-西德利公司（英国）
hazard and risk analysis	危害与风险分析
hazard identification and vulnerability assessment	危害确认和漏洞评估
hazard in case of failure	失效时的危害性
hazard mitigation program	减灾计划
hazard prevention methodology	危害预防方法
head of the flame tube	火焰筒头部
head resistance	迎面阻力
head-contractor	总承包商
Headquarters of General Staff of Chinese People's Liberation Army	中国人民解放军（原）总参谋部
head-up display	（驾驶员面前的）平视显示器
head-up guidance system	（飞机着陆）平视导航系统
health & safety	健康与安全
health monitoring	健康监测
health safety & environment	健康安全与环境
hearing range	听觉范围
heat conduction	热传导

heat convection	热对流
heat dissipation	热耗散
heat energy	热能
heat energy added	增加热能
heat exchanger	热交换器
heat expansion coefficient	热膨胀系数
heat flow rate	热流率
heat flux density	热流密度
heat management system	热处理系统
heat radiation	热辐射
heat rate	热耗率
heat recovery	热回收
heat recovery steam generator	热回收蒸汽发生器
heat release rate	热释放率
heat resistance	热阻
heat resistant alloy	耐热合金
heat resistant cast steel	耐热铸钢
heat sealing strength	热合强度
heat shield	隔热罩
heat shock resistance	抗热冲击
heat transfer	传热学
heat treat condition	热处理条件
heat treatable alloy	可热处理合金
heat treatment furnace	热处理炉
heating and pressurizing inlet air endurance test	进气加温加压持久试验
heavy load turbine	高负荷涡轮
heavy-lifting copter	重型直升机
heavyweight power turbine	重型动力涡轮
height above touchdown	距着陆区高度
height overall	全高
helical compressor	螺旋压气机
helical gear	螺旋齿轮、斜齿轮
helical spline	螺旋键槽

helical spline coupling	螺旋键槽联轴器
helical spring	螺旋弹簧
helical spur gear	斜齿圆柱齿轮
helicopter autorotation	直升机自旋
helicopter rotor head	直升机旋翼桨毂
heliplane	(新型)旋翼飞机
helium	氦气
helmet mounted display	头盔显示器
Hemisphere	半球-公务机(塞斯纳飞机公司,美国)
Hero's engine	赫罗的发动机
Herschel-Quincke tubes	赫歇尔-昆克管
Hertz	赫兹
hexagon socket screw	内六角螺钉
hexagonal head bolt	六角头螺栓
Hexel	(美国)赫克塞尔复合材料公司
hierarchical control	分层管控
high alloy steel	高质合金钢
high alpha research vehicle	(美国国家航空航天局)高攻角实验机
high aspect ratio	大展弦比
high aspect ratio wing	大展弦比机翼
high capacity twin aisle	大型双通道飞机
high carbon steel	高碳钢
high concealing	高隐蔽性
high cycle fatigue	高周疲劳
high cycle fatigue life	高周疲劳寿命
high degree of sweep	高度后掠
high development risk	高开发风险
high effectiveness advanced turbine	高效先进涡轮
high efficiency deep grinding	高效深度磨削
high end tools	高级工具
high energy density beam processing	高能束流加工技术
high energy ignition box	高能点火控制器
high energy ignition system	高能点火系统

high energy ignition units	高能点火装置
high entry barrier technology	高壁垒技术
high entry cost	高成本进入市场
high even tail	高平尾
high expansion ratios	高膨胀比
high fidelity system	高保真系统
high flow solid titanium fan	大流量实心钛合金风扇
high frequency noise	高频噪声
high idle	高速慢车
high idle rating	高速慢车推力
high intensity radiated fields	高强度辐射性场
high lift blading	高升力叶片
high lubricity	高润滑性
high Mach turbine engine	高马赫数涡轮发动机
high mass capture ratio	高流量系数
high octane fuels	高辛烷值燃油
high power slam accelerate	大功率迅速加速
high precision	精度高
high precision dynamic strain measurement	高精度动态应力测量
high pressure	高压
high pressure cock	高压开关
high pressure combustion test	高压燃烧试验
high pressure compressor	高压压气机
high pressure compressor delivery temperature	高压压气机出气温度
high pressure compressor disc load	高压压气机轮盘载荷
high pressure compressor entry pressure	高压压气机入口压力
high pressure compressor entry temperature	高压压气机进气温度
high pressure compressor exit pressure	高压压气机出口压力
high pressure compressor guide vane leading edge chamfer	高压压气机导向叶片前沿导角
high pressure compressor guider throat	高压压气机导向器喉道
high pressure compressor stall	高压压气机失速
high pressure compressor surge margin test	高压压气机喘振裕度试验
high pressure drive shaft	（启动用）高压驱动轴

high pressure shaft governor	高压轴调节器
high pressure spool seizure	高压轴受卡
high pressure turbine	高压涡轮
high pressure turbine efficiency	高压涡轮效率
high pressure rotor bow	（停车后因受热不均而产生的变形）高压转子弯曲
high pressure turbine rotor inlet	高压涡轮转子进气
high pressure turbine stage loading	高压涡轮级负荷
hail ingestion test	吸雹试验
high pressure shut-off valve	高压节流阀
high pressure turbine	高压涡轮
high pressure valve	高压阀
high pressure vessel	高压容器
high reliability coating	高可靠性涂层
high risk manoeuvre	高风险动作
high sensitivity to notches	对凹槽高度敏感
high shear rivet	高剪力铆钉
high speed civil transport	高速民用运输机
high speed power turbine	高速动力涡轮
high speed research	（美国国家航空航天局）高速研究计划
high speed research-enabling propulsion materials	（美国国家航空航天局）高速研究计划-高效推进材料
high speed research-propulsion	（美国国家航空航天局）高速推进研究计划
high stability engine control	高稳定性发动机控制
high strain rate	高应变率
high stress level	高应力
high temperature air heat exchanger	高温空气散热器
high temperature demonstration unit	高温验证机
high temperature measurement	高温测量
high temperature turbine test	高温涡轮试验
high tension	高张力
high throughput	高通量

high value target	高值目标
high velocity jet efflux	高速喷气流
high velocity oxygen fuel	高速火焰（喷涂）
high velocity oxygen fuel coating	高速火焰涂层
high velocity oxygen fuel thermal spray	高速火焰热喷涂
high-altitude long-endurance	高空长航时
high-angle-of-attack technology program	高攻角技术项目
high-capacity high hot strength liquidspray combustion	高容热强度液雾燃烧
high-end business jet	高级公务机
higher humidity in the cabin	较高的座舱湿度
higher quality officer	（英国国防部）高级质量官员
highest production rate	最高产量
high-growth industry	高增长工业
high-heating value	高热值
highlight	亮点
highly customer oriented	以客户为中心
highly efficient embedded turbine engine	（美国）高效埋入式涡轮发动机
highly fretting resistant	高抗磨损
highly instrumented	装有大量测试装置
highly integrated digital electronic control	高度集成数字式电控
highly loaded bowed stator	高负荷弯曲静叶
highly skewed propeller	高倾斜螺旋桨
highly specialized manufacturing technology	高度专业化生产技术
highly stressed part	高应力部件
high-margin spare parts	高利润零备件
high-speed acquisition	高速采集
high-speed milling machine	高速铣床
high-speed milling of blisk	高速铣削整体叶盘
high-speed punching machine	高速冲床
high-speed turbine engine demonstrator	（美国）高速涡轮发动机验证机（计划）
hight throughout	高流量
hight throughout	高产出
hightech tanker	高科技空中加油机

high-temperature material	高温材料
high-temperature fuel cell	高温燃料电池
high-temperature resins	高温树脂
high-temperature reusable surface insulation	（美国国家航空航天局）高温重复使用表面绝热材料
high-temperature vacuum brazing	高温真空钎焊
high-temperature，high-strength structural support foam	（美国开发的）高温高强结构支撑泡沫材料
high-thrust engine	大推力发动机
high-turning bowed compressor stator	大折转角弯曲压气机静叶
high-value patent	高价值专利
hijack-proof piloting system	防劫机领航系统
Hindustan Aeronautics Ltd.	（印度）锡都斯坦航空公司
hinge joint	铰链连接
hingeless rotor	无绞旋翼
hire-purchase	分期付款购买
Hispano-Suiza	（法国赛峰集团子公司）希斯巴诺—苏伊莎
histogram	柱状图
historical reference collection	历史性参考文献汇编
hold open rod	保持打开支撑杆
holding company	控股公司
holding down bolt	固定螺栓
holding pattern	待命航线
hold-your-breath time	屏住呼吸时间
holes and grooves	孔和凹槽
hollow aluminium alloy fan	空心铝合金风扇叶片
hollow revet	空心铆钉
hollow titanium alloy fan	空心钛合金风扇叶片
hollow-bladed blisk	空心整体叶盘
Honda Jet	本田-公务机(本田公司,日本)
honeycomb core	蜂窝芯
honeycomb sandwich structure	蜂窝夹层结构
honeycomb seal	窝状封严篦齿

Honeywell International	(美国)霍尼韦尔国际
Hong Kong Aero Engine Services Limited	香港航空发动机服务公司
hoop stress	周向应力
horizontal audit	水平审计(同级别间审计)
horizontal broaching machine	卧式拉床
Hornet	大黄蜂-战斗机(麦道公司,美国)
horsepower	马力
horsepower limiter	马力限制器
horseshoe vortex	马蹄涡
hot air anti-ice system	热空气防冰装置
hot air manifold	热空气集气环
hot and high requirements	高温高原环境的要求
hot calibration wind tunnel	热校准风洞
hot day take off	高气温起飞
hot dip galvanized	热镀锌
hot forging	热锻造
hot forming	热成形
hot hubbing	热模压
hot isostatic press-casting	等压热烧结模压铸件
hot rolled bar steel	热轧钢条
hot start	热开车
hot test	热试车
hot turbine discharge gases	高温涡轮出口燃气流
hot-day take-off	高温起飞
hot-end parts	热端部件
hot-selling aircraft	热销飞机
hot-shot ignition	热射流点火
hovering ceiling	悬停升限
hovering flight	悬停飞行
hovering in ground effect	地面效应悬停
hovering out of ground effect	离地悬停
hovering out of ground effect ceiling	离地悬停升限
hub and spoke construction	轮辐结构
hub contour	轮毂外形

hub/tip ratio	轮毂比
human error	人为差错
human hearing sensitivity	人类听觉敏感度
human-centred design	以人为本的设计
humidity	湿度
hundredweight	英担
Hurel-Hispano	宇雷尔-伊斯帕诺公司(赛峰集团,法国)
hush kit	降噪强制混合器
hybrid digital/analog computer system	数/模混合计算机系统
hybrid fan vectored thrust engine	混合风扇推力矢量发动机
hybrid laminar-flow control	混合式层流控制
hybrid linings	混合式减噪声
hybrid propulsion system	混合推进系统
hybrid-grid computational fluid dynamics simulation	流体动力学混合网格模拟
hybrid-mount	混合式吊装
hydraulic	水力的
hydraulic amplifier	液压放大器
hydraulic brake power-measuring device	水力测功器
hydraulic control unit	液压控制器
hydraulic delay unit	液压延迟器
hydraulic nozzle	液压喷嘴
hydraulic oil indicator	液压油指示器
hydraulic oil sensor	液压油传感器
hydraulic press	水压机,液压锻造设备
hydraulic pump	液压泵
hydraulic pump failure	液压泵失效
hydraulic seal	液压封严件
hydraulic servo driving	液压伺服驱动
hydraulic servo power unit	液压伺服马达
hydraulic starter	液压起动机
hydraulic vibration table	液压振动台

hydrocarbon scramjet engine technology	用碳氢化合物燃料的冲压发动机技术
hydrocarbons	碳氢化合物
hydrocooling filter	液冷过滤器
hydrofoil	水翼
hydrogen	氢气
hydrogen relief treatment	去氢处理
hydrolic forming	液压成形
hydromechanical backup	以液压机械系统为备份
hydromechanical control system	液压机械式控制系统
hydro-mechanical unit	液压机械组件
hydrometer	比重计
hydroplane	水上飞机
hyperburner	超燃加力器
hypersonic aerothermodynamic theory	高超声速气动热力理论
hypersonic flight	高超声速飞行
hypersonic inlet	高超声速进气道

Ice cell	冰室
ice extraction	除冰
ice ingestion	吞冰
ice shedding rubber tip	防冰橡胶尖部
ice slab ingestion gun	吸冰块试验用炮
ice slab ingestion test	吸冰块试验
icing awareness	结冻探测
icing fog	冰雾状
icing spraybar system	喷冰系统
icing tunnel test	冰风洞试验
idle	慢车
idle descent power	慢车下降功率
idle power	慢车功率

idle reverse	慢车反推
idle run	慢车
idle thrust	慢车推力
idle valve	慢车活门
idler gear drive	惰轮齿轮驱动
idler gearshaft	惰轮齿轮轴
idling rating	慢车状态
idling valve	慢车活门
igniter coil	点火线圈
igniter lead	点火器导线
igniter plug	点火塞
ignition	点火
ignition altitude	点火高度
ignition exciter	点火励磁器
ignition harness/lead	点火线
ignition limit	点火极限
ignition period	着火周期
ignition point	燃点
ignition relay	点火继电器
ignition temperature	点燃温度
ignition test	点火试验
ignitor	点火器
ignitor lead	点火器电缆
ignitor plug	点火器火花塞
illustrated parts catalogue	零部件图解目录
Ilyushin Aviation Complex（Ilyushin）	（俄罗斯）伊留申航空集团
imaging sensor equipment	成像传感器设备
imaging techniques	成像技术
Israeli Military Industries	（以色列）军事工业公司
imide oligomers	（高温高性能）酰亚胺低聚物
impact by bird ingestion	鸟吸影响
impact resistant	抗冲击
impact resistant thermoplastics	抗冲击热塑性塑料
impact-rub	碰摩

impeller	叶轮
impeller shaft	叶轮轴
impeller wheel	叶轮
imperfection	瑕疵
imperfection parts	有瑕疵部件
impingement cooling	气流冲击冷却
impingement film floatwall	冲击膜浮壁
impingement plate	冲击板
impingement tube	冲击冷却管
implementation of modifications on engine in service	对在役发动机实施改装
implementation of the design change	设计更改的实施
implementation of the new components	新部件的生产实施
implementation potential	实施潜力
implementation strategy	实施策略
improved digital electronic engine controller	改进的数字电子式发动机控制器
improved turbine engine program	(美国陆军)改进涡轮发动机计划
improving process performance	提高流程能力
impulse turbine	脉冲式涡轮
impulse-reaction combination	冲动-反动组合
impulse-reaction turbine	冲动-反动涡轮
impulse-type turbine	冲击式涡轮
impulsive noise	脉冲噪声
in process quality control	制程中的质量管理
in service date（s）	在役天数
inability to shut the engine down	失去停车能力
inadvertent thrust deployment	无意中打开反推
inaugural flight	首航
incapacitate crew or passenger	失去行为能力的机组或乘客
incidence angle	迎角
incidence of failure	故障发生率
inclined combustion chamber	倾斜燃烧室
inclusion	夹杂
inco713LC blade	材料为镍合金(inco713LC)叶片

inco718 welded drum with inco blading	带镍合金叶片的镍合金材料（inco718）焊接盘鼓
income statement	收益表
income taxes	所得税
incoming quality control	进料质量管理
Inconel	（一种耐高温镍合金钢）因科耐尔
independent final and process gauging	独立的最终和中间测量
independent overspeed protection	独立超速保护
indicated air speed	实测相对飞行速度
indicated altitude	仪表指示高度
indicated Mach number	实测马赫数
indicated mean effective pressure	显示的平均有效压力
indicated power	（仪表）指示功率
indication	（无损探伤检测）示像
indigenous defense fighter	（台湾）经国号战斗机
indirect operating costs	间接运营成本
indirect purchasing control specifications	间接采购控制规定
individual income tax	个人所得税
inducer	进口导风轮
inductance discharge type	电感放电式
induction coil	感应线圈
induction motor	感应电机
induction system icing	进气系统结冰
induction system icing test	进气系统结冰试验
Industria de Turbo Propulsores	（西班牙）涡轮发动机工业公司
industrial endoscope	工业内窥镜
industrial rigid borescope	工业硬性窥镜
inertia	惯性
inertia friction welding	惯性摩擦焊
inertia load	惯性力
inertial engine development program	初始发动机研发项目
inertial friction welding	惯性摩擦焊
inertial measurement unit	惯性测量系统
inertial navigation	惯性导航

inertial navigation system	惯性导航系统
in-flight air quality	飞行中座舱空气质量
in-flight engine failure	发动机空中失效
in-flight entertainment	机上娱乐
in-flight inadvertent reverser deployment	飞行中无意识使用反推装置
in-flight power loss rate	空中功率损失率
in-flight recorder	飞行记录器
in-flight shut down	空中停车
in-flight shut down rate	空中停车率
info update	信息更新
Information Office of the State Council	国务院新闻办公室
information technology	信息技术
infrared countermeasures	红外对抗
infrared detection set	红外探测仪
infrared pyrometer	红外高温计
infrared radiation	红外辐射
infrared radiation source of aircraft	飞机红外辐射源
infrared search and track	红外线搜索跟踪系统
infrared starting thermal imaging system	红外凝视热像仪
infrared stealth technology of aircraft	飞机红外隐身技术
ingestion capacity	吸入能力
ingestion resistance	抗外物击伤能力
ingot mold	铸锭模
ingots	铸锭
inherent boundary layer transition control	固有附面层过渡控制
initial crack size	初始裂纹尺寸
initial flaw size	初始缺陷尺寸
initial maintenance inspection test	初始维修检查试验
initial public offering	原始股公开发行
initial swirl	初始漩流
injection molding	注塑成型
injection molding machine	注塑机
injection pump	引射泵
injector	喷嘴

inlet flow fields	进气道流场
inlet additional drag	进气附加阻力
inlet aerodynamics	进气空气动力学
inlet airflow distortion test	进气畸变试验
inlet anti-icing	进气道防冰
inlet boundary layer	进口附面层
inlet buzz	进气喘振
inlet bypass door	进气道放气门
inlet characteristics	进气特性
inlet cone	进气锥
inlet cowl	进气道
inlet distortion	进气畸变
inlet duct doors	进气道放气门
inlet duct rumble	进气道隆隆声
inlet dynamic response	进气动态响应
inlet dynamic distortion	进气动态畸变
inlet external drag	进气外阻
inlet flow field	进气道流场
inlet guide vane	进气导向叶片
inlet kinetic energy efficiency	进气动能效率
inlet lip	进气道唇口
inlet mass flow ratio	进气流量系数
inlet operating condition	进气工作状态
inlet outer barrel	进气道外筒
inlet particle separator	进气道颗粒分离器
inlet pressure check valve	进气压力检查阀
inlet pressure recovery	进气压力恢复
inlet pressure recovery coefficient	进气压力恢复系数
inlet spray intercooling	进气道喷雾级间冷却
inlet stability margin	进气稳定裕度
inlet starting	进气起动
inlet swirl flow distortion	进气旋流畸变
inlet throat	进气喉道
inlet total pressure distortion	进气总压畸变

inlet total pressure recovery	进气总压恢复
inlet total temperature distortion	进气总温畸变
inlet/engine compatibility	进气道-发动机相容性
inlet-distortion simulating screen	进气畸变模拟网格
inner V Blade	内 V 形叶片
inner air casing	内空气机匣
inner annulus	火焰筒内环
inner barrel	内筒
inner cone	内锥
inner crack	内裂纹
inner defect	内缺陷
inner diameter	内径
inner fixed structure	内部固定结构
inner gas path	内气道
inner loop thrust vectoring	内环推力矢量
inner nose fairing	内锥整流罩
inner reduction gearbox	体内减速器
inner shroud	内盖
innovative technologies	创新技术
innovative cavity concepts	创新的空腔方案
innovative future air transport system	(欧盟)创新型未来空运系统
inorganic high-temperature paints	无机物高温涂漆
inorganic-organic nanocomposite	有机-无机纳米复合材料
in-production engines	在生产的发动机
insert tongues correctly removed	嵌入的舌片已被正确拆除
inserts	榫头、镶嵌件
inserts correctly seated	正确安装插入装置
in-service engine	在役发动机
in-service recorders	服役记录器
in-service simulation	服役模拟
insider trading	内部交易
inspection certificate	检验证书
inspection history record card	检验历史记录卡
inspection procedure card	检验流程卡

installation	安装
installation constraint	吊装限制条件
installation kit	安装工具
installation losses	安装损失
installation memorandum	发动机安装备忘录
installation thrust	（已考虑安装损失的）净推力
installed	已安装的
installed specific fuel consumption	装机后的耗油率
instantaneous sound pressure	瞬时声压
instantaneous status	瞬态参数
instrument flight rules	靠仪表盲飞规则
instrument landing system	仪表着陆系统
instrument panel	仪表板
instrumentation standards and techniques	测试标准和技术
insulating blanket	隔热层
insulating sleeve	绝缘套
insulation resistance	绝缘电阻
intake	进气道
intake chutes	进气槽
intake guide vanes	进口导流叶片
intake louvre	进气道放气口
intangible assets	无形资产
integral bearing	整体式轴承
integral drive arm	整体式驱动臂
integral drum	整体鼓筒
integral heat transfer rib	一体化的导热肋板
integral inference technology	一体化预测技术
integral ring	整体环
integrally bladed ring	整体叶片环
integrally bladed rotor	整体叶片转子
integrated actuation package	综合电动液压作动器
integrated airport management	综合机场管理
integrated assessment	综合评估
integrated blade moment weighing	叶片力矩的综合称量

integrated business	综合业务
integrated certification schedule	一体化取证时间表
integrated coast guard system	联合海岸防卫系统
integrated control and safety system	综合控制和安全系统
integrated design system	一体化设计系统
integrated drive generator	组合驱动发电机
integrated drive generator surface air cooled oil cooler	组合驱动发电机的表面气冷滑油冷却器
integrated electric drive	整体式电驱动
integrated engine management system	综合发动机管理系统
integrated fairing	一体化整流罩
integrated fuel pump and control	一体化燃油泵与控制系统
integrated gasification combined-cycle	一体化的气化联合循环
integrated high performance turbine engine technology	(美国)高性能涡轮发动机综合技术项目
integrated logistics centre	综合性后勤中心
integrated maintenance information system	综合维护信息系统
integrated modular avionics	综合模块式航电
integrated power system	一体化的动力系统
integrated product development，production and maintenance	一体化产品开发、生产和维护
integrated product team	综合产品开发组
integrated programme team	综合项目小组
integrated project	综合项目
integrated project management team	综合项目管理团队
integrated propulsion control	推进系统综合控制
integrated propulsion control system	一体化推进控制系统
integrated propulsion system	一体化推进系统
integrated propulsion，lift and control	动力、增升和控制一体化
integrated resilient aircraft control	一体化弹性飞机控制
integrated standby flight display	一体化待机飞行显示
integrated standby instrument system	综合应急仪表系统(GE)
integrated starter and generator	一体化的启动机和发电机
integrated starter generator	一体化的起动器和发电机

integrated suite of computer aided design	计算机辅助设计综合程序包
integrated supply chain	一体化的供应链
integrated voice communication system	飞行器通信系统
integration capability（robust control of sub-tiers）	集成能力（对子供应商的有力控制）
integration definition document	一体化方案文件
integration requirements document	一体化需求文件
integration studies	综合研究
intellectual property rights	知识产权
intelligent engine system	智能发动机系统
intelligent fault-tolerant control	智能容错控制
intelligent flight control system	智能飞机控制系统
intelligent sensors	智能传感器
intelligent transportation system	智能交通
intensive flying trials unit	（英国）强化试飞队
intentional unbalance	有意不平衡
inter turbine burning	（美国）内燃涡轮技术
interactive data management platform	交互式数据管理平台
interactive electronic technical manual	交互式电子技术手册
intercase strut	中介机匣支柱
interceptor	截击机
interchangeability	互换性
interchangeability code	互换性代号
interconnecting piping	连接管路
intercontinental traffic	洲际交通
intercooled	级间冷却
intercooled cycle	间冷循环
intercooled recuperated	间冷回热
intercooled recuperated aero-engine	间冷回热航空发动机
intercooler	间冷器
intercooling	级间冷却
interdisciplinary research centre	跨学科研究中心
interest expense	利息费用
interface control document	接口控制文件

interference drag	干涉阻力
interlocked blade	预扭自锁叶片
interlocking flap	互锁板
intermediate casing	中介机匣
intermediate gearbox	中间齿轮箱
intermediate pressure	中压
intermediate pressure compressor	中压压气机
intermediate pressure power offtake	中压转子功率提取
intermediate pressure turbine	中压涡轮
intermediate rated power	中等额定功率
intermediate rating	中等推力
intermediate reduction gearbox	中间减速器
intermediate stage	中间级
intermediate support	中介支承
intermediate zone	过渡区
intermetallic compound	金属间化合物
internal air system	内部空气系统
internal classification	内部分类
internal combustion engine	内燃机
internal compression inlet	内压式进气
internal fluid simulation	（叶片）内流模拟
internal gearbox radial driving shaft	机内齿轮箱径向传动轴
internal rate of return	内部盈利率
internal shaft	内轴
internal stress	内应力
International Aero Engines	国际航空发动机公司
International Aero Engines supplier	国际航空发动机公司的供应商
International Aerospace Quality Group	国际航空航天质量协会
international air transport agreement	国际航空运输协议
International Air Transport Association	国际航空运输协会
International Air Transport code number	国际航空运输协会编号
international business	国际业务
international circulation permit	国际通行许可证
International Civil Aviation Organization	国际民航组织

International Civil Aviation Organization Assembly	国际民航组织大会
International Civil Aviation Organization certification landing/take-off cycle	国际民航组织着陆/起飞循环取证标准
International Civil Aviation Organization Committee on Aviation Environmental Protection-8	国际民航组织航空环境保护委员会第 8 级标准
International Civil Aviation Organization limit for post 1996 engine	1996 年起实施的国际民航组织发动机限定值
International Congress of Aeronautical Sciences	航空科学国际大会
International Field Office（FAA）	（美国联邦航空局)外航业务办公室
International Field Unit（FAA）	（美国联邦航空局)外航业务处
international financial reporting standards	国际财务报告标准
international knot	国际节
international laboratory accreditation cooperation	国际实验室认证协作机构
International Leasing and Finance Corporation	国际金融租赁公司
International Monetary Fund	国际货币基金组织
International Organisation for Standardisation	国际标准化组织
international register of certificated auditors	有资质的审核员国际注册
International Space Station	国际空间站
international space station task force	国际空间站任务组
international standard atmosphere	国际标准大气
international standard atmosphere conditions	国际标准大气条件
International Standard for Environmental Management System	国际标准化环境管理体系
international standard for phytosanitary measures	国际植物检疫标准
International Standard for Quality Management System	国际标准化质量管理体系
International Standards Organisation	国际标准化组织
international traffic in arms regulations	（美国)武器国际贸易条例
internet security system	互联网安全系统

interpretive explanation material	(适航要求)说明性解释资料
intersecting hole	交叉孔
intershaft bearing	中介轴承
intershaft hydraulic seal	轴间液压封严
intershaft support	中介支承
interstage bleed	级间放气
interstage bleed valve	级间放气阀
interstage brush seals	级间刷式密封
interstage honeycomb seal	级间窝状封严
interstage parameter measurement	级间参数测量
interstage seal	级间封严
Interstate Aviation Committee	俄罗斯适航局
Interstate Aviation Committee Aviation Register	俄罗斯适航局航空记录
inter-turbine temperature	级间涡轮温度
interval and space ratio of signals	信号时间间隔和占空比
in-time inventory	零库存
invar	殷瓦钢(不膨胀钢)
inventories	库存
inverse boundary element method	反向边界元法
inverse bow	反向弧
inverted exhaust turbofan	倒置排气涡扇发动机
inverted velocity profile	倒置速度截面
investment in enterprise	对企业投资
investment portfolio	投资证券组合
investments	投资
investments and services	投资和服务
invitation to tender	招标
invoice query notification	账单查询通知
ion beam machining	离子束加工
Irkut Corporation (Irkut)	(俄罗斯)伊尔库特公司
iron bird	铁鸟试验台
irrotational flows	无旋流动
isentropic efficiency	等熵效率

isentropic exponent	等熵指数
isentropic flow equation	等熵流动方程
Ishikawajima Harima Heavy Industries	(日本)石川岛播磨重工业公司
Isograph	(英国开发的)可靠性分析软件
isolated nacelle line	发动机短舱外形线
isolation control unit	隔离控制系统
isolation control valve	隔离控制阀
isolation threshold value	隔离门阈值
iso-propyl-nitrate starting	异丙基硝酸酯起动
isostatic presse	等静压机
Israel Aerospace Industries	以色列航空航天工业公司
Israeli Military Industries	以色列军事工业公司
iteration	迭代
iterative process	迭代流程
Ivchenko-Progress Design Bureau (Ivchenko-Progress)	(乌克兰)伊夫琴科-进步设计局
izod impact strength	悬臂梁冲击强度

Japan Aerospace Exploration Agency	日本航空航天开发机构
Japanese Aero Engine Corporation	日本航空发动机公司
Jeffcott rotor-stator systems	杰夫科特转子-定子系统
jet	喷气
jet efflux	喷流
jet exhaust excess impulse	排气过剩冲量
jet flaps	喷气襟翼
jet fuel	喷气发动机燃料
jet hardening	喷流硬化
jet in crossflow through primary hole	主燃孔横向射流
jet noise	喷气噪声
jet orifice	喷油孔
jet pipe	喷管

jet pipe overhaul	尾喷管大修
jet pipe temperature	喷管温度
jet quenching	喷流淬火
jet-baffle amplifier	喷嘴挡板式放大器
jet-flap blade	喷气襟翼叶片
Jethete	杰赫不锈钢
Jethete billets	杰赫不锈钢锻坯
Jethete forging	锻件杰赫不锈钢
Jethete rings	杰赫不锈钢环
jetpipe	喷管
Jindivik	金迪维克-靶机(阿姆斯特朗-西德利公司,英国)
Jirnov engine	季尔诺夫发动机
Jirnov vortex gas turbine engine	季尔诺夫涡流燃气涡轮发动机
joint airworthiness requirement	欧洲联合适航要求
joint army-navy specifications (USA)	美国陆军和海军通用规范
Joint Aviation Authorities	欧洲联合航空局
joint aviation authorities transition	欧洲联合航空局过渡性机构
joint aviation requirement-engine	欧洲联合适航对发动机的要求
joint aviation requirements	欧洲联合航空规范
joint cargo aircraft	(美国陆军和空军)联合运输机
joint definition bulletin	联合定义公告
joint definition phase	联合定义阶段
joint design standard	联合设计标准
joint engineering procedure	联合工程程序
joint engineering specifications	联合工程规范
joint engineering standard	联合工程标准
joint expendable turbine engine concept	(美国)一次性涡轮发动机综合方案
joint multi-role rotorcraft	联合多用途旋翼机
joint or unique	联合还是单干
joint program office	综合计划办公室
Joint Strike Fighter (Lightning Ⅱ)	(美国)联合攻击机(闪电Ⅱ)
joint tactical information distribution system	联合战术信息分布系统
joint technical meeting	联席技术会议

joint technology demonstrator engine	综合技术验证发动机
joint technology initiative	(欧盟项目)联合技术计划
joint venture	合资
joint venture company	合资公司
joint-unmanned combat air vehicle	联合无人战斗机
Joule	焦耳
Joule-process	焦耳-过程
journal oil shuttle valve	油梭往复阀
junction box	接线盒
just in time production	零库存生产
just-in-time	准时

kaizen	看板管理
kaizen methodology	改善方法
Kawasaki Heavy Industries	(日本)川崎重工公司
kerosine	煤油
Kevlar	凯夫拉(一种高强度玻璃纤维环氧树脂)
Kevlar wrapped containment	用凯夫拉材料缠绕的包容机匣
key business figures	关键业务数字
key characteristics	主要特性
key performance indicator	主要性能显示器
key ranges	主要航程
kilojoule	千焦耳
kilometre	千米
kilometric fuel consumption per seat	座公里耗油量
kilometric fuel consumption per ton payload	吨公里商载耗油量
kilopascal	千帕斯卡
kilowatt	千瓦
kinematic viscosity	运动黏度
kinetic air pump	动力空气泵

kinetic energy	动能
kinetic metallization	动态金属涂层
kinetic valve type amplifier	喷射式放大器
King Air	空中国王-公务机（比奇飞机公司，美国）
Kingdom of Saudi Arabia Presidency of Civil Aviation	沙特阿拉伯王国民航局
kits	组件
Klimov Design Bureau (Klimov)	（俄罗斯）克里莫夫发动机设计局
Klimov Public Joint-Stock Company	（俄罗斯）克里莫夫联合股份公司
Klimov's vectoring thrust	克里莫夫推力矢量
knock	爆震
knots	节（速度）
knowledge based engineering	知识工程
knowledge-based development processes	以知识为基础的开发流程
Korea Aerospace Industries Ltd.	韩国航空工业公司
Krestrel	红隼-战斗机（霍克-西德利公司，英国）
Kuznetsov Scientific and Technical Complex	（俄罗斯）库兹涅佐夫科研生产联合体
Kyoto mechanisms	京都机制
Kyoto Protocol	（减排）京都协定书

labour-intensive	劳动密集型
labour-intensive enterprise	劳动密集型企业
labyrinth assembly	篦齿封严组件
labyrinth seals	篦齿式封严件
laissez-faire approach	（商贸的）自由放任主义
lamilloy	多孔层板结构
laminar plasma spraying	层流等离子喷涂
laminar profile	层流翼型

laminar restrictor	层流限流器
laminar separation	层流分离
laminar-flow engineering	层流工程
laminated flame tube wall	层板结构火焰筒壁
laminated plastic panel	层压塑料板
laminated restrictor	层板节流器
laminated sheet cooling blade	层板冷却叶片
lam-turn transition	层流向紊流转变
Lancer	枪骑兵-轰炸机(波音公司,美国)
land-based test site	陆上试验台
landing and take off	着陆和起飞
landing at maximum landing weight	以最大着陆重量着陆
landing configuration	着陆状态
landing distance	着陆距离
landing distance available	可用的着陆滑跑距离
landing fees	着陆费
landing field elevation	飞机起降机场海拔高度
landing flare	着陆前拉平飞机
landing gear	起落架
landing gear bay	起落架舱
landing gear control handle	起落架操纵手柄
landing gear safety pin	起落架保险销
landing gear shock absorber	起落架减震器
landing gear support beam	起落架支撑梁
landing slots	着陆空位
lapse rate	温度递减率
large airliner	大型客机
large bird ingestion test	大鸟吸入试验
large business jets	大型公务喷气机
large cargo freighter	大型货机
large eddy simulation	大涡模拟
large freight door	大型货舱门
large freighter	大型货机
large scale wind turbine generator	大型风力发电机

large-eddy simulation	大涡模拟
large-scale advanced propfan	先进桨扇发动机大型验证机
largest-ever firm engine order	有史以来最大的发动机订单
Larzac	拉扎克发动机（赛峰集团，法国）
laser beam	激光束
laser beam machining	激光束加工
laser cladding	激光熔覆
laser cladding technology	激光熔覆技术
laser cutting	激光切割
laser Doppler velocimeter	激光多普勒测速仪
laser drilling	激光制孔
laser drilling of contoured hole	激光仿形钻孔
laser flash method	激光闪射法
laser formation	激光成形
laser gradient rig test	激光梯度台架试验
laser inspection	激光检测
laser marking and carving	激光打标与雕刻
laser peening technology	激光喷丸技术
laser prototyping	激光快速成型
laser remelted ceramic coatings	激光熔覆陶瓷涂层
laser shock peening	激光喷丸
laser shock processing	激光冲击处理
laser surface treatment	激光表面处理
laser tracker	激光跟踪仪
laser ultrasonic visualizing inspector	激光超声波可视化检测仪
laser welding	激光焊接
laser-glazed coating	激光表面熔覆涂层
last article inspection report	末件产品检验报告
latch access panel	带有门闩的检修面板
latest component technologies	最新部件技术
lathing center	车削中心
lattice	点阵、格状
lattice design	点阵结构设计
lattice imaging	晶格成像

lattice sizing optimisation	点阵结构体积优化
lattice structure	点阵结构
launch	启动
launch customer	启动客户
Laval nozzle	拉瓦尔喷口
layered manufacturing technology	分层制造技术
LCA Tejas	光辉-战斗机（印度斯坦航空公司，印度）
lead time	交货时间
leading edge	前缘
leading edge aviation propulsion	（CFM 公司）先进航空推进技术项目
leading edge film cooling	前缘气膜冷却
leading edge technology	前沿技术
leading supplier	先进供应商
leading-edge composites expertise	先进的复合材料专业技术
leading-edge flap	前缘襟翼
leading-edge slat	前缘缝翼
lead-times	制造周期
leaf seal	薄片封严
leakage flows	泄流
lean assembly	精益装配
lean blade	倾斜叶片
lean burn combustor	贫油燃烧室
lean development	精益发展
lean limit	贫油极限
lean manufacturing	精益生产
lean principles	精益理念
lean staged combustion	贫油分级燃烧
lean year	精益年
lean-premixed-prevaporised combustion	贫油预混气相燃烧
lean-staged combustor	贫油分级燃烧室
LEAP56	（CFM 公司）CFM56 发动机技术更新计划
Learjet	利尔-公务机（塞斯纳飞机公司，美国）

learning management system	学习管理系统
lease engines	租用发动机
least material condition	最少留料情况
least mean square	最小均方法
left hand	左手
left hand side	左手侧
Legacy	莱格赛-商务飞机(安博威公司,巴西)
legal assets	法定资产
legal entity	法人,法律实体
legal finding	法律问题
legal names of contract parties	合同方的法定名称
leisure airline	旅游航空公司
length overall	全长
length/diameter	长/径比
less complex design	更简单的设计
less-polluting jet	低污染喷气飞机
letter of credit	信用证
letter of intend	意向书
level gauge	液位计
leveraged buy-out	杠杆收购
leveraged leasing	杠杆租赁
liabilities	负债
liabilities & equity	负债和权益
liberty works	自由工厂(罗罗)
lier one supplier	一级供应商
life and rotor integrity strategy	寿命和转子一体化战略
life cycle cost	全寿命周期费用
life discipline manual	寿命规范手册
life limited part	有限寿命零件
life limited part cost	有限寿命零件费用
life management plan	寿命管理计划
life management process	寿命管理程序
life of critical components	关键零部件寿命
life technique	定寿方法

lifetime protection	全寿命保护
lifetime support	全寿命支持
lift component effects	升力部件影响
lift dumpers	减升板
lift fan	升力风扇
lift/drag ratio	升阻比
lifting analysis	升力分析
lifting fan	升力风扇
lifting fixtures	吊架
lifting line	升力线
light around	联焰
light combat helicopter	轻型战斗直升机
light-emitting diode	发光二极管
light fabricated	轻结构的
Light Helicopter Turbine Engine Company	(罗罗-霍尼韦尔合资)轻型直升机涡轮发动机公司
light maintenance	简单维护
light observation helicopter	轻型侦察直升机
light single-engined aircraft	轻型单引擎飞机
light sources	光源
light sport aircraft	轻型运动飞机
light transmittance	透光率
light turbulence	弱湍流
light utility helicopter	轻型通用直升机
light-duty gas turbine	轻型燃机
lightning strike	闪电
lightweight fan case	重量轻的风扇机匣
line cost	航线工时费
line labour	航线检修工时
line maintenance	航线维护
line maintenance part	航线维护件
line matrix core	线基核心
line replaceable item	外场可换零件
line replaceable module	航线可换模块

line replaceable unit	航线可更换设备
line replacement	航线更换
linear acoustic liner in intake	网状进气道隔音层
linear extrapolation	线性外插法
linear friction welding	线性摩擦焊
linear interpolation	线性内插法
linear liner	线性衬垫
linear motor	直线电机
linear variable differential transformer	线性可调差动变压器
linearised Euler method	线性化欧拉方法
liner C263	材料为 C263 的内衬
line-side support	场外服务
lipped ring design	唇形环设计
liquefied natural gas	液化天然气
liquid crystal display	液晶显示器
liquid flowmeter calibrator	液体流量表校准设备
liquid fuel	液体燃油
liquid metal embrittlement	液态金属脆化
liquid molding technology	液体成型及预型件技术
liquid oxygen	液氧
liquid phase infiltration joint	液相渗透连接
liquid rocket motor	液体火箭发动机
liquid water content	液态水含量
liquidated damages clause	清算损失条款
liquidized petrolic gas	液化石油气
liquid-spray combustion	液雾燃烧
liquified hydrogen	液化氢气
listed company	上市公司
littoral combat ships	(美国)濒海战舰
load bearing	承力轴承
load deflection test	负荷挠度试验
load factor	(客机的)上座率
load factor	(飞机)过载系数
load limit and ultimate	载荷极限和破损极限

load simulator	载荷模拟试验台
load spectrum	载荷谱
load supporting system	承力系统
load transferring scheme in static structure	静力结构传力方案
load-bearing part	承载部件
load-bearing structure	承载结构
loading	载荷
loading of aircraft accessory	飞机附件载荷
loading slot	承载槽
loading spectrum	载荷谱
loan capital	借贷资本
loan guarantee	贷款担保
lobe-type nozzle	瓣形喷管
local sourcing	本地采购
locating pin	定位销
location specific property	位置特性
lock-bolt	锁紧螺栓
Lockheed Martin Corporation	(美国)洛克希德-马丁公司
locking device	锁定装置
locking nut	紧固螺帽
locking screw	锁紧螺钉
lock-nut	锁紧螺帽
lock-pin	锁销
lock-plate	锁片
lockup device	锁死装置
logistics provider	后勤保障供应商
long development time	长周期开发
long duct nacelle	长涵道短舱
long haul flight	长航程飞机
long on-wing time	长时间在役
long range	远程
long-haul destination	远距离目的地
Longitude	经度-公务机(塞斯纳飞机公司,美国)

long-lead drawing	长周期图纸
long-range aircraft	长航程飞机
long-range cruise speed	远程巡航速度
long-term agreement	长期合作协议
long-term debt	长期负债
long-term investment	长期投资
long-term service agreement	长期售后服务协议
loose or missing part	松弛的部件或缺少零件
loose part	松弛的部件
Los Angeles International Airport	洛杉矶国际机场
loss coefficient	损失系数
loss of stability pressure ratio	稳定压比损失
loss of tail rotor effectiveness	尾桨失效
loss of thrust	推力损失
loss-making	亏损
lost wax casting	脱蜡铸造
lost wax molding	脱蜡造模法
lost-time accident	造成误工的事故
lot number	批号
lot reject rate	批退率
Lotarev	洛塔列夫发动机(进步设计局/ZVL, 乌克兰/捷克)
low alloy steel	低级合金钢
low aspect ratio	低展弦比
low aspect ratio blading	低展弦比叶片
low cost aircraft leasing	低成本飞机租赁
low cost economies	低成本经济
low cost of ownership	低购置成本
low cost region implants	在低成本地区建厂
low cycle and high cycle compound fatigue	高低周复合疲劳
low cycle fatigue	低周疲劳
low cycle fatigue life	低周疲劳寿命
low density materials	低密度材料
low direct operation cost	低运营成本

low drag	低阻力
low emissions alternative power	（美国国家航空航天局）低污染备选动力计划
low energy	低能量
low energy air	低能空气
low frequency noise	低频噪声
low hub diameter fan	低轮毂直径风扇
low idle	慢车
low noise aircraft	低噪声飞机
low octane fuels	低辛烷值燃油
low oil pressure sensor	低油压传感器
low pilot workload	降低飞行员工作负荷
low pitch	低螺距
low pressure	低压
low pressure combustion test	低压燃烧试验
low pressure compressor	低压压气机
low pressure cooler	低压冷却器
low pressure fuel valve	低压燃料活门
low pressure pump	低压泵
low pressure rotor drive device	低压转子传动装置
low pressure shaft governor	低压轴调节器
low pressure shut-off valve	低压燃油切断阀
low pressure system	低压系统
low pressure turbine	低压涡轮
low rate initial production	小批初始生产
low refurbishment cost	低成本翻修
low Reynolds number flow	低雷诺数气流
low risk design philosophy	低风险设计原则
low speed power turbine	低速动力涡轮
low speed stall	低速失速
low spool turbomachinery component	低压涡轮机部件
low stage check valve	低级压气部分检测阀
low stress no distortion	低应力无变形
low stress non-deformation	低应力无形变

low stress non-deformation welding	低应力无形变焊接技术
low tension	低张力
low value-add	低附加值
low-carbon economy	低碳经济
low-end products	低端产品
lower bifurcation nose fairing	低分离整流罩
lower control limit	控制下限
lower efflux	低位喷流
lower heating value fuel	较低热值燃料
lower specification limit	规格下限
lower thrust settings	较低推力设置
low-fare air travel	廉价空中旅行
low-loss design	低损耗设计
low-rate initial production	少量首批生产
low-temperature fuel cell	低温燃料电池
low pressure shaft speed	低压转子转速
low pressure spool seizure	低压轴受卡
low pressure turbine	低压涡轮
low pressure turbine efficiency	低压涡轮效率
low pressure turbine exit pressure	低压涡轮出口压力
low pressure turbine rotor exit	低压涡轮转子排气
low pressure turbine rotor inlet	低压涡轮转子进气
LS-DYNA	（ANSYS 公司的）非线性有限元分析软件
low stress no distortion welding	低应力无变形焊接
Lubbock fuel nozzle	拉伯克燃料喷嘴（可调进口燃料喷嘴）
lube coefficient	滑油参数
lube oil console	滑油箱架
lubricant	润滑剂
lubricating oil	润滑油
lubricating oil cooler	滑油冷却器
lubricating oil pump	滑油泵
lubricating system	润滑系统
lubricating system interface	润滑系统界面

lubricating system performance	润滑系统性能
lubrication and scavenge oil pump	润滑与回油泵
Luftfahrt Forschung Program	(德国)航空研究计划
Luftfahrt-Bundesamt	德国民航局
Lufthansa Technics Shenzhen	深圳汉莎技术有限公司
Lufthansa Technik	汉莎技术公司

Mach number	马赫数
machine to shape	机加成形
machined graphite mould	机加石墨模
machined surface	机加工表面
machining center	加机中心
machining cooled airfoils	气冷叶片的机加工
machining equipment	机加设备
macro analysis	宏观分析
magnesium intercase	镁材料中介机匣
magnesium matrix composite	镁金属基复合材料
magnet systems technology	磁力系统技术
magnetic amplifier	磁放大器
magnetic chip detector	磁塞铁屑探测器
magnetic crack detection	磁力裂纹探测
magnetic flux	磁通量
magnetic lens	磁镜
magnetic particle inspection	磁粉检测
magnovalve	被动温控流量阀(罗罗专利)
maiden flight	首飞
main beam	吊挂主承力梁
main burner	主燃烧室
main driving shaft	主传动轴
main engine controller	发动机主控制器
main gear box	主齿轮箱

main mixing inner port	主混区内端口
main mixing outer port	主混区外端口
main moulding box	主铸模盒
main oil pressure sensor	主滑油管压力传感器
main oil temperature sensor	主滑油管温度传感器
main orifice	主孔
main reduction gearbox	主减速器
main rotor blade	主旋翼桨叶
main rotor gearbox	主旋翼减速器
main supply route	主要供货渠道
main tank	主油箱
main transmission system	主传动系统
maintainability	维护性
maintainability design	维修性设计
maintenance analysis	维修分析
maintenance and engineering management	维修与工程管理
maintenance cost	维护费用
maintenance cost index	维修费用系数
maintenance cost per hour	每小时维修成本
maintenance cost per hour agreement	按小时付费维护协议
maintenance cost per hour programs	按小时付费维护方案
maintenance data acquisition unit	维护数据采集装置
maintenance depot	维修站
maintenance error decision aid	维修差错决断法（波音）
maintenance error management system	维修差错管理系统
maintenance flaw and error	维修疏忽和差错
maintenance hanger	飞机维修库
maintenance manual	维护手册
maintenance objective	维修日标
maintenance organisation exposition （EASA Part 145）	维修组织说明（欧洲航空安全局 145 部）
maintenance organization approval	维修许可证
maintenance planning guide	维修规划指南
maintenance precautions	维护注意事项

maintenance repair and overhaul	维护、维修和大修
maintenance review board	维修审核委员会
maintenance review board report	维修审查委员会报告书
maintenance standard	维修标准
maintenance statement	维修规范
maintenance steering group	维修指导组
maintenance-by-the-hour	计时维修
major aerodynamic rig test	重要气动部件试验
major disruption	严重损害
major quality investigation（Rolls-Royce）	重要质量调查（罗罗）
make to print	接图制造
make vs buy	自制与采购的比较
make-buy strategy	自己制造或外部采购战略
make-versus-buy selection	自制与外购对比选择
MAKO	马可-教练机（欧洲航宇防务公司，欧洲合作）
malfunction of the engine	发动机故障
mainline shaft	主轴
malsynchronisation	不同步
managed service	直接参与的服务
management by objectives	目标管理
management fee drag	管理费
management information system	管理信息系统
management of non-conformance	无超差管理
management of the modification flow	改造流程管理
management shake-up	管理层人员变更
mandatory target	强制性目标
maneuver control system	操纵控制系统
maneuver load	机动载荷
maneuvering characteristics augmentation system	操纵特性强化系统
manoeuvrability	机动性
manoeuvre envelope	机动性包络
manometer	液体压力表
manual change request	手册变更要求

manual draft approval	手册初审
manual drive unit	手动装置
manual single-phase valve	手动单向活门
manual switching	手动转换
manual trim	手动配平
manual tuning	手动调谐
manual's life	手册寿命
manufacture order	生产单
manufacturing acceptance sheet	生产许可证
manufacturing alteration memorandum	生产变更备忘录
manufacturing alteration request	生产变更请求
manufacturing bill of materials	制造物料清单
manufacturing change request	生产变更申请
manufacturing critical part plan	关键部件生产计划
manufacturing cycle	制造周期
manufacturing engineering	制造工程
manufacturing flexibility	制造适应性
manufacturing instruction	工艺说明
manufacturing jigs and tooling	制造用模具和工装
manufacturing laboratories catalogue	工艺实验室目录
manufacturing launch plan	生产启动计划
manufacturing process instruction	初始生产工艺
manufacturing responsible party company	负责生产制造的伙伴公司
manufacturing sensitive part plan	灵敏部件生产计划
manufacturing technical package	制造技术工作包
manufacturing technique	制造技术
manufacturing technique instruction	制造技术规定（罗罗）
manufacturing tolerance	制造公差
manufactury technique	制造技术
margin	裕度
marine application	船舶应用
marine total care	船舶产品包修协议（罗罗）
marine trent engine	舰船用遄达发动机
marine turbine	船用涡轮

maritime patrol aircraft	海上巡逻机
maritime reconnaissance aircraft	海上侦察机
maritime search and rescue aircraft	海上搜索救援飞机
market player	市场参与者
market scenario	市场风险
market value added	市场增加值
Marubeni Corporation	(日本)丸红株式会社
mass flow	流量
mass flow function	流量函数
mass flow rate	质量流率
mass transfer	质量传输
Massachusetts Institute of Technology	麻省理工学院
master alloy	母合金
master die	标准模
master parts list	主要零件表
master program	总体计划
master stree-number of cycles curve	(应力与循环次数的关系)主 S - N 曲线
master switch	主开关
material and process specifications	材料和工艺规范
material delamination	材料分层
material for trial orders	首批材料订单
material increase manufacturing	增材制造
material inspection specification	材料检验规范
material release order	材料放行单
material requirements/resource planning	材料要求和资源计划
material return authorisation	退货审批
material return instruction	退货通知
material review application	材料应用审核
material review board	材料审核委员会
material specification	材料规范
material specification Rolls-Royce	罗罗公司材料规范
material test methods	材料测试方法

materials & mechanical methods	（罗罗公司设计手册）材料和结构计算方法
materials laboratory services	材料实验室的服务
materials safety data sheets.	材料安全数据表
materiel acceptance and release form	物料收发单（罗罗）
mating jig	组装台
mating surface	配合面
mature labour cost	成熟期劳工成本
mature maintenance cost	成熟期维护成本
mature product	成熟产品
mature shop labour	成熟期维修工时
mature shop material	成熟期维修材料
Maxflyer	（一种近太空飞行器）"最高飞行者"
maximum allowable weight	最大允许重量
maximum bending moment	最大弯矩
maximum bending stress	最大弯应力
maximum ceiling	最大升限
maximum climb	最大爬升
maximum continuous	最大连续
maximum continuous rating	最大连续状态
maximum cruise rating	最大巡航推力状态
maximum cyclic life	最大循环寿命
maximum declared values	最大申明值
maximum design landing weight	最大设计着落重量
maximum dry specific fuel consumption	中间状态耗油率
maximum empty weight	最大空重
maximum endurance	最大续航时间
maximum fuel capacity	最大燃油载荷
maximum instantaneous distortion	最大瞬时畸变
maximum landing weight	最大着陆重量
maximum level speed	最大平飞速度
maximum material condition	最大实体尺寸
maximum normal working pressure	最大正常工作压力
maximum payload	最大商载

maximum ramp weight	最大滑行重量
maximum reverse	最大反向
maximum side load	最大侧向载荷
maximum take off weight	最大起飞重量
maximum take-off	最大起飞
maximum take-off thrust	最大起飞推力
maximum take-off weight	最大起飞重量
maximum zero-fuel weight	最大无油重量
Mazak machining centre	(马扎克公司)马扎克机加工中心
McDonnell Douglas	(美国)麦克唐纳-道格拉斯公司
mean R/S diffusion factor	平均动/静叶扩压因子
mean stress	平均应力
mean time between failures	平均故障间隔时间
mean time between failures	平均无故障时间
mean time between maintenance	平均维修间隔时间
mean time between overhaul	平均翻修寿命
mean time between removal	平均拆卸间隔时间
mean time between unscheduled removal	平均非计划拆换间隔时间
mean time to failure	平均无故障时间
mean time to repair	平均维修时间
meanline analysis	中线分析
meantime to failure	平均故障前时间
measurement and control system for engine test	发动机试验测控系统
measurement system evaluation	测试系统评估
measures of performance	绩效考评标准
mechanical	机械的
mechanical design	机械设计
mechanical drive application	机械驱动应用
mechanical drive device	机械驱动装置
mechanical failure	机械故障
mechanical impedance	机械阻抗
mechanical integrity	机械完整性
mechanical power offtake	输出机械功率

mechanical rig programme	部件结构试验
mechanical risk	结构风险
mechanical shock test	机械冲击试验
mechanical speed	机械速度
mechanical torque multiplier	机械式扭矩放大器
mech-insensitive structural stress method	网格不敏感结构应力计算方法
medium business jet	中型公务喷气机
medium multi-role combat aircraft	中型多用途作战飞机
medium-single	中型单通道飞机
megajoule	兆焦耳
megawatt-scale	兆瓦级
melting crucible	熔炉
membrane-assistant resin infusion	薄膜辅助液态成型技术(奥地利 FACC 公司)
memorandum of understanding	谅解备忘录
memorandum of deposit	存放单(存款单)
memorandum of policy	政策大纲
Mercure	水星-客机(达索公司,法国)
mercury	汞柱
merger and acquisition	兼并与收购
meridional view	子午线视图
Messier-Bugatti	(法国航空产品制造商)梅西埃-比加蒂公司
metal arc spray	金属电弧喷涂
metal brush seal	金属刷密封
metal coating	金属涂层
metal fatigue	金属疲劳
metal matrix composite	金属基复合材料
metal nanoparticles burning	金属纳米粒子的燃烧
metal powder	金属粉末
metal screen knitting	金属网编织
metal vapour	金属蒸气
metallic coating	金属涂层
metallic extruded finned tube	金属轧片管

metallic flexible laminated membrance coupling	金属叠片挠性联轴器
metallized solid fuel	金属化固态燃料
metal-oxide coatings	金属氧化物涂层
methanol engine	甲醇发动机
method of manufacture	制造方法
Metropolitan Washington Airports Authority	美国华盛顿市区机场管理局
micro air vehicle	微型无人飞行器
micro crack	微裂纹
micro electro mechanical systems	微电动机械系统
micro structure	微结构
micro turbine/microturbine	微型燃气涡轮机
microelectronics	微电子
microengine technology	微型发动机技术
micro-positioning mechanism	微定位机构
micropulse propeller	微型脉冲推力器
micro-sensors	微传感器
microswitch on engine	发动机微动电门
microthruster	微型推进器
Microturbine	微型涡轮发动机公司
micro-vortex generator	微型涡流发生器
microwave sensing technology	微波探测技术
mid power reslam	中功率迅速再加速
mid size airliner	中型客机
mid size freighter	中型货机
mid tandem fan	中置风扇
mid turbine frame	中置涡轮框架
mid-air refuelling aircraft	空中加油机
middle of the market	中间市场概念(波音)
middle of the market aircraft	中间市场客机(波音)
midsize long-haul jetliner	中型长航程喷气式客机
mid-span radius	中跨半径
mid-span shroud contact surface	叶片凸肩接触面
mid-span shroud of blade	叶片中间凸肩
mid-span shroud strength	叶片凸肩强度

mid-span shrouded blade	带凸肩的风扇
mid-thrust geared turbofan engine	中等推力齿轮转动涡扇发动机
Mikoyan Design Bureau	米高扬设计局
miles per hour	英里每小时
milestone	里程碑
military spending	军费开支
military standard（USA）	（美国）军队标准
millimetre	毫米
mini borescope	微型内窥镜
mini combustor	微小型燃烧室
minimal quantities of lubricant	最少切削液
minimum augmentation rating	最小加力状态
minimum constant speed	最小恒定转速
minimum customer bleed pressure	最低用户输出气压
minimum drag speed	最小助力速度
minimum engine	最低性能发动机
minimum equipment list	最低设备放行
minimum flow valve	最小油量活门
minimum fuel pressure limiter	最小燃油压力限制器
minimum ignition energy	最小点火能
minimum level of safety	最低安全水平
Ministero della Difensa（Italian MoD）	（意大利）国防部
Ministry of Defence（UK）	（英国）国防部
ministry of defence quality assurance representative	（英国）国防部质保代表
ministry of information industry	（中国）信息产业部
ministry of public security	（中国）公安部
minority interests	少数股东权益
Mirach	（意大利）"米拉奇"靶机
Mirage	幻影-战斗机（达索公司，法国）
mirror milling system	蒙皮镜像铣切系统
miscellaneous	杂项
miscellaneous publication/brochure	各种出版物和小册子
mismatch	装配错误

misoriented grains	杂晶
mission adaptive wing	任务自适应机翼
mission and reserve rule	飞行任务和余量规定
mission fuel contingency	余油量
mission ready management solution	完成任务管理方案
mission ready management solutions	(罗罗公司军用发动机)备勤管理方案
missioncare agreement	发动机任务保障协议(罗罗)
mistake proofing	检验错误
mistuned bladed disc	非谐调叶盘
Mitsubishi Aircraft Corporation	(日本)三菱飞机制造公司
Mitsubishi Heavy Industries	(日本)三菱重工业公司
Mitsubishi regional jet	三菱支线客机(三菱重工,日本)
mixed compression inlet	混压式进气道
mixed electronic-hydromechanical control system	电子-液压机械混合式控制系统
mixed flow vectored thrust	混合流矢量推力
mixed jet exhaust system	混合流喷气系统
mixed oxides of nitrogen	混合氮氧化物
mixed reality	混合现实技术
mixed-flow compressor	混流压气机
mixer	混合器
mixer chute	混合器斜槽
mixing noise	混合噪声
mixing of alternate definition	综合定义
mixing tube	混合筒
mobi-ticket	移动电子客票
mock up	木制样机
modal analysis	模态分析
modal parameter	模态参数
model based definition	基于模型的设计思想
model of Langston	(流体)郎斯通模型
model specification for military engines	军用发动机型号规范
model verification	模型检验
moderate stage loading	适中级负荷

modern aero engine industries	现代航空发动机工业
modern working practices	现代工作实施
modification	修改
modification & design approval	重大设计改装批准书
modification standard	修正标准
modular affordable product line	模块化成本可负担产品线
modular design	单元体设计
module exposure	单元体拆卸
module strip	单元体分解
moisture content	含水量
mold to shape	浇模成形
molecular weight	分子量
Mollier diagram	莫里尔图
molten carbonate fuel cell	熔融碳酸盐燃料电池
moment drag	动量阻力
moment weighing of blades	叶片的力矩动量
monetary policy tools	货币政策工具
money-saver aircraft	省钱飞机
monoball bearing housed in intercase splitter	单球轴承置于中介机匣的隔离架中
monocrystal blades	单晶叶片
monohull	单体船
monolithic titanium	整块钛
more electric aircraft	多电化飞机
more electric engine	多电化发动机
more intelligent engine	更加智能化的发动机
morphing aircraft	(美国)变形体飞机
most sensitive hearing range	最敏感的听觉范围
Motor Sich JSC	马达西奇发动机公司(乌克兰)
Motoren Und Turbinen Union Munchen GmbH	(原称慕尼黑发动机涡轮联合公司)摩天宇航空发动机公司(德国)
motor-glider	动力滑翔机
mould support chill	铸模支柱激冷
mount	安装,安装节
mount configuration	安装方式

mount design criterion	安装结构设计准则
mount static load	安装结构静载荷
mount stressing	安装强度分析
mounted pod installation	吊舱式安装
mounting	(发动机)安装架
mounting flange	安装法兰
moving assembly line	移动总组线
moving target indicator	移动目标显示器
moving wave	动波
Mriya	梦幻-运输机（安东诺夫公司,乌克兰）
MTU Aero Engines GmbH	(德国)摩天宇航空发动机公司
MTU Aero Engines GmbH-Development and Design Center	摩天宇航空发动机公司-开发和设计中心
MTU Aero Engines North America Inc.	摩天宇航空发动机北美公司
MTU Maintenance Berlin-Brandenburg GmbH	摩天宇维修柏林-勃兰登堡公司
MTU Maintenance Canada Ltd.	摩天宇维修加拿大有限公司
MTU Maintenance Hannover GmbH	摩天宇维修汉诺威公司
MTU Maintenance Zhuhai Co. Ltd.	摩天宇维修珠海有限责任公司
MTU Turbomeca Rolls-Royce GmbH	摩天宇透博梅卡罗罗公司
MTU Turbomeca Rolls-Royce，Munich	摩天宇透博梅卡罗罗公司,慕尼黑
multi-role transport	(德国)多用途运输机
multi-axis grinding	多轴磨床
multi-axis thrust vectoring	多轴推力矢量
multi-blade propeller	多叶螺旋桨
multi-cylinder engine	多缸发动机
multi-disciplinary optimized design	多学科优化设计
multi-engine gear drive	多台发动机共轴齿轮传动
multi-frequency tuning	多频调谐
multi-fuel capability	使用多种燃料的能力
multi-function display	多功能显示器
Multifunctional Frontline Fighter	(俄罗斯)多用途前线战斗机计划
multi-grains	多晶

multi-manning	多技能
multi-mission maritime aircraft	多用途海上飞机
multi-pass solenoid valve	多通路电磁阀
Multiple Application Propfan Study	桨扇发动机多重应用研究
multiple combustion chamber	多个燃烧室
multiple pure tone	多重纯声
multiple spools	多轴（发动机）
multi-rib integral panel structure	多肋整体壁板结构
multi-role combat aircraft	多用途战斗机
multi-role stealth aircraft	多用途隐形飞机
Multi-Role Tanker Transport	（德国）多用途加油运输机
multi-scale eddy structure	多尺度涡结构
multi-service stealth fighter	多用途隐形战斗机
multi-shaft layout	多轴布局
multi-spool compressor	多转子压气机
multi-way discharge nozzle	多路灭火喷管
Mustang	野马-公务机（塞斯纳飞机公司，美国）
mutually beneficial partnership	互利互惠伙伴关系

nacelle	发动机短舱
nacelle air-spaces	短舱的风道
nacelle anti-ice	短舱防冰
nacelle design criteria	短舱设计规范
nacelle lead time	短舱研制周期
nacelle line spec freeze	短舱外形定型
nacelle zone ventilation	短舱的通风
nagative incidence stall	负迎角失速
naked resistance	净阻力
nano air vehicle	超微型无人飞行器
nanocomposite	纳米复合材料
nanoflake sensors	纳米薄片传感器

nanometer level	纳米级
narrow strip peen forming	窄条喷丸成型
narrowbody	窄体飞机
NASA John C. Stennis Space Center	美国国家航空宇航局斯坦尼斯航天中心（美国）
NASA mini-engine	美国国家航空航天局的微型发动机
NASA morphing airplane	美国国家航空航天局的变形飞机
（NASA）Propulsion Systems Laboratory	（美国国家航空航天局）推进系统实验室
National Accreditation Bodies	国家认证机构
National Advisory Committee for Aeronautics	（美国航空宇航局的前身）国家航空咨询委员会
National Aeronautic Association	（美国）全国航空协会
National Aeronautics and Space Administration	美国国家航空航天局
National Aerospace and Defence Contractors Accreditation Programme	世界航空航天和国防业务承包商认证机构
National Aerospace Industry Associations	全国航空航天工业联合会
national aero-space plane	国家航天计划
national aerospace standards	国家航空航天标准
national air and space museum	国家航空航天博物馆
National Air Transportation Association	（美国）全国航空运输联盟
national airworthiness authority representative	国家适航当局代表
National Archives and Records Administration	国家档案与记录管理局
National Audit Office of the People's Republic of China	中华人民共和国审计署
National Aviation Facilities Experimental Center	国家航空设施实验中心
National Bureau of Statistics	（中国）国家统计局
National Business Aviation Association	（美国）全国公务航空联盟
National Codification Bureau	（北约）国家编码局
national explosive-proof regulation	国家防爆设计规范
national flights	国内航班
national full-scale aerodynamics complex	国家全尺寸空气动力学综合设施
National Institute of Aerospace	国家航空宇航研究所

National Institute Standards and Technology	(美国)国家标准与技术研究院
national metrology standard	国家计量标准
National Quality Assurance Authority	(英国)国家质量保证局
national quality assurance representative(s)	国家质量保证代表
National Research Council	美国国家研究委员会
National Transport Safety Bureau	美国国家运输安全署
National Transportation Safety Board	国家运输安全委员会
national vehicle and fuel emissions laboratory	国家飞行器与燃油排放实验室
national voluntary laboratory accreditation program	(美国)国家实验室义务认证计划
NATO Electronic Warfare Advisory Committee	北约电子战咨询委员会
NATO Eurofighter and Tornado Management Agency	北约欧洲战斗机和狂风战斗机管理局
NATO Eurofighter Management Organisation	北约欧洲战斗机管理机构
NATO frigate helicopter	北约护卫舰载直升机
NATO integrated extended air defense system	北约一体化扩展防空系统
NATO stock number	北约(13位数)物料编号
natural gas solenoid valve	天然气电磁阀
natural laminar flow	自然层流
nautical mile	海里
Navier-Stokes equations	纳维—斯托克斯方程
navigation fees	导航费
Navy of Chinese People's Liberation Army	中国人民解放军海军
near field noise	近场噪声
negatively scarped intake	下唇伸出的进气道
Nene engines	尼恩发动机(罗罗公司,英国)
neon	氖气
nested loop	嵌套循环迭代法
net assets	净资产
net fixed assets	固定资产净值
net income	净收入
net income after interests and taxes	息税后净收入
net income after taxes	税后净收入
net income before interest and taxes	息税前收入

net profit	净利润
net sales	净销售额
net sales revenue	销售收入净值
net thermal efficiency	净热效率
net thrust	净推力
net-shape manufacturing	净形制造
network manufacturing	网络化制造
network safety management	网络安全管理
network system control	网络化系统控制
neural network	神经网络
NEURON	神经元-无人机(达索公司等,西欧多国)
new aero engine core concepts	(欧洲)新型航空发动机核心机方案
new aircraft platforms	新的飞机平台
new engine concepts	近的发动机方案
new engine option	换发方案(空客)
new factory in greenfield site	绿地上的新工厂
new maintenance concepts	新型维护方法
new midsize aircraft	新型中型客机
new production introduction	引入新产品
new repair processes	新的修理工艺
new small aircraft	新型支线机项目(波音)
New York Stock Exchange	纽约证券交易所
new-generation engine	新一代发动机
Newton	牛顿
Newton metre	牛顿米
Newtonian fluids	牛顿流体
Nexcelle	(短舱)奈赛公司
next generation air transportation system	(美国)下一代航空运输系统
next generation flight management	新一代的飞行管理
next generation launch technology	(美国国家航空航天局)下一代发射技术
next higher assembly	次高组合件

Nextant	莱斯顿-公务机（莱斯顿航空宇航公司，美国）
Ni-based	镍基合金
Ni-based cast	镍基合金铸造
Ni-based forged	镍基合金锻造
nick	刻痕，裂口
nickel aluminium	铝化镍
nickel aluminium wire	镍铝丝
nickel cast iron	镍铸铁
nickel casting	镍基铸件
nickel chrome steel	镍铬钢
nickel chromium wire	镍铬丝
nickel high-temperature alloy	镍金高温合成
nickel steel	镍钢
night flight ban	夜间禁飞
Night Intruder	"夜侵者"无人机
night target observation	夜间目标观察
night vision goggles	航空夜视镜
Nighthawk	夜莺-战斗机（洛克希德-马丁公司，美国）
Nimrod	猎迷-反潜机（英宇航公司，英国）
nitric oxide	一氧化氮
nitriding	渗氮
nitrogen	氮气
nitrogen generation system	氮气发生器
nitrogen oxides	氮氧化物
nitrogen peroxide	二氧化氮
no fragment	无碎片
no fuselage	无机身
no life-limited part	无限寿命部件
no on-wing inspection	无须在翼检查
no purchase order no payment	不下采购单，不付款
no swarf	没有细屑
no unauthorised change	没有未经许可的变更

no weight distribution valve	无重力输油阀
noble metal heat treatment	贵金属材料的热处理
noise abatement	减噪
noise abatement take-off technique	减噪起飞方法
noise absorbent material	吸音材料
noise absorber panels	吸声板
noise attenuation	噪声衰减
noise criteria	噪音准则
noise exposure forecast contour	飞机噪音预测等量线
noise footprint	噪声影响区
noise monitoring terminal	噪音监察站
noise prediction	噪声预估
noise reduction technologies	降噪声技术
noise reflective surface	噪声反射面
noise suppression installation	消音器安装
noise-sensitive airports	低噪声机场
noise-suppression	噪声抑制
noise-suppression nozzle	消声喷管
nomenclature	命名法
Nomex	(杜邦公司生产的)芳香族聚酰胺耐高温纤维
nominal cone	标称圆锥
nominal engine capability	发动机标定能力
nominal gap	标称间隙
nominal power	额定功率
nominal pressure	公称压力
nominal strain at fracture	断裂标称应变
nominal stress	标称应力
non destructive examination	无损检验
non destructive testing	无损检测
non intrusive stress measurement system	非插入式应力测量系统(普惠)
non mission capable supply	(军机)停机待件
non return valve	单向阀
non volatile memory	非易失性内存

non-acceptable imperfection	不可接受的瑕疵
non-adherence to procedures	不遵守规范
non-aero derivative	非航空衍生产品
non-axisymmetric nozzle	非轴对称喷管
non-binding agreement	无约束性协议
non-burn titanium	阻燃钛合金
non-cash assets	非现金资产
non-chromated technology	无铬酸盐技术
non-compliances	不合格
non-compliances profile	不合格简介
non-compliances report	不合格报告
non-conformance authority	受权处理异常零件的人或组织
non-conformance issue management	异常零件问题处理
non-conformance report	异常零件报告
non-contact multi-directional magnetising disc testing	无接触多向磁化盘检验
non-critical parts	非关键性部件
non-current assets	非流动资产
non-destructive inspection	无损检测
non-destructive testing	无损损伤检验
non-destructive testing equipment	无损探伤设备
non-destructive testing specification	无损检测规范
non-ferrous metals	有色金属
non-flame type ignition device	非火焰形点火装置
non-imperfection product management	产品无瑕疵管理
non-intersection tolerance of three axis	三轴不相交度
nonlinear dynamics	非线性动力学
nonlinear transient calculation	非线性瞬态计算
nonmetallic coating	非金属涂层
non-operating revenue	营业外净收入
non-perpendicularity tolerance of three axis	三轴不垂直度
non-predictable failure	不可预测失效
non-premium traffic	(航空公司的)普通舱运量
non-pressure welding	无压焊接

non-productive assets	非生产性资产
non-recurring cost	非经常性成本
non-recurring engineering	非经常性工程
non-recurring engineering cost	非经常性工程成本
non-relevant indication	假象
non-return valve	单向阀
non-routine activities	非例行活动
non-standard equipment	非标准设备
nonstationary stochastic process and signal processing	非平稳随机过程和信号处理
nonsteady aerodynamics	非定常气动力学
nonvacuum electron beam welder	非真空电子束焊机
non-value added time	无效时间
normal cruise	正常巡航
normal lapse rate	正常递减率
normal start	正常起动
Norsk Jet Motors	(挪威)喷气发动机公司
North Atlantic Treaty Organisation	北约组织
Northrop-Grumman Corporation	(美国)诺斯罗普-格鲁曼集团
nose cone	中心锥
nose cone housing	前锥机匣
nose-down pitch	下俯
nose-up pitch	上仰
not applicable	不适用
not otherwise specified	除非另有指定
notch sensitivity	缺口敏感性
notice for changing forecast	更改预估量的通知
notice of proposed rulemaking	建议标准通告
notice of proposed rulemaking-stage 5	建议第五阶段标准
notice to suppliers	给供应商的通知
notification of acceptance for application	受理申请通知书
notional engine concepts	概念发动机方案
not-to-exceed ceiling price	不可超过的最高价
novel construction	新结构

novel fuel injection for mixing enhancement in gas turbine combustors	改进燃气轮机燃烧室混合性的新型燃料喷注技术
nozzle	喷管
nozzle acoustic test rig	尾喷声学台架试验
nozzle base drag	喷管底阻
nozzle cast CI023	材料为 CI023 的铸造喷嘴
nozzle expansion ratio	喷管膨胀比
nozzle flow calculation	喷管流动计算
nozzle flow coefficient	喷管流量系数
nozzle flow fields	喷管流场
nozzle fully open	喷口完全打开
nozzle fully shut	喷口完全关闭
nozzle gas dynamics	喷管气动力学
nozzle guide vane	导向器叶片
nozzle oil pump	喷口滑油泵
nozzle operating ram	喷口作动筒
nozzle operating roller	喷口收放滚棒
nozzle operating sleeve	喷口操纵套管
nozzle orifice	喷嘴孔口
nozzle pressure ratio	喷管压比
nozzle thrust coefficient	喷管推力系数
nozzle thrust gain	喷管推力增益
nozzle velocity coefficient	喷管速度系数
NPO Saturn JSC（Saturn）	（俄罗斯）土星科研生产联合体股份公司
nuclear acceptance standard	核工业产品验收标准
nuclear aero-engine	核能航空发动机
nuclear energy for the propulsion of aircraft	核能飞机动力
nuclear manufacturing services	核工业生产制造服务
numerical calculation of incompressible flows	不可压流数值计算
numerical calculations of two-phase flows	两相流数值计算
numerical control	数字化控制
numerical control lathe	数控车床

numerical modelling method	数值建模方法
numerical propulsion system simulation	数字推进系统仿真
nut	螺母

Oasis	(杜邦公司生产的)复合电气绝缘材料
objective life	实际寿命
oblique shock	斜激波
oblique wing	斜翼
oblique-wing transport	斜机翼运输机
occupational health	职业健康
occupational health and safety assessment series（ISO）	职业健康与安全管理体系
off limit stress levels	超出范围的应力级
off wing operations room	地面运营中心
off-design engine performance	非设计点发动机性能
off-design point	非设计点
off-gas combustor	废气燃烧室
Office National d'Etudes et Recherches Aérospatiales	(法国)国家航空航天研究院
office of advanced research and technology	先进研究与技术办公室
office of aeronautics and space technology	航空宇航技术办公室
offset package	补偿贸易工作包
offtake	提取
off-the-shelf	现货供应
off-the-shelf equipment	现成设备
off-wing maintenance	(发动机)拆离飞机的维修
oil cock	滑油开关
oil consumption	滑油消耗量
oil control module	滑油控制单元
oil damped bearing	挤压油膜轴承
oil debris monitoring system	滑油碎屑监控系统

oil differential pressure switch	滑油压差开关
oil feed pump	滑油泵
oil film clearance	油膜间隙
oil film stiffness	油膜刚度
oil filter	滑油过滤器
oil filter debris analysis	滑油过滤器碎屑分析
oil filter differential pressure sensor	滑油过滤器压差传感器
oil flow interruption	滑油流量中断
oil heat exchanger	滑油热交换器
oil heat rejection	滑油放热
oil level indicator	油面指示器
oil level meter	滑油油量表
oil level sensor	滑油位传感器
oil pressure gauge	滑油压力表
oil pressure indicator	滑油压力表
oil pressure limit	滑油压力极限
oil pressure transmitter	滑油压力传感器
oil pump	滑油泵
oil pump pack	滑油泵组件
oil pump unit	滑油泵
oil quantity indicating system	滑油油量指示系统
oil rotating seal	油封严
oil scavenging	回油
oil starvation	缺油
oil sump pressure	滑油内腔压力
oil tank	滑油箱
oil tank access door	油箱门
oil temp sensor	滑油温度传感器
oil temperature indicator	滑油温度指示表
oil temperature limit	滑油温度极限
oil temperature transmitter	滑油温度传感器
oil vent	滑油气孔
oil way	油路
oil/fuel temp/pressure switch	滑油、燃油温度和压力测量开关

oil/oil cooler	用滑油为冷媒的滑油冷却器
oil-free rotor support	无油转子支承
Olympus	奥林普斯发动机(罗罗/斯奈克马公司,英国/法国)
omnibearing ion implantation and deposition	全方位离子注入与沉积
omnirange radio beacon	全向式无线电信标
Omsk Engine Design Bureau	(俄罗斯)鄂木斯克发动机设计局
on site support	现场支持
on time delivery	按时交付
on wing support	在役支援
onboard amenity	机上服务设施
on-condition maintenance	视情维护
one engine inoperative	单发失效
one-dimensional steady channel	一维定常管流
on-engine condition monitoring	发动机自备状态监控
one-off charge	一次性付款
one-off engine inspection	发动机一次性检查
one-piece composite fuselage	整体复合材料机身
one-stop convenience	一站式便利
one-way clutch	单向离合器
Oneworld Alliance	环宇一家
online aerospace supplier information system	在线航空航天供应商信息查询系统
on-line communications	实时沟通
on-line customer web center	在线客户网络支持
online exchange	在线交换
OnPoint solutions	发动机全面服务方案(GE)
on-site	现场
on-site failure warning	现场故障报警
on-site reps	现场代表
on-time handover	准时交付
on-wing care	在翼维护
on-wing maintenance	在役维护
open architecture	开放式结构
open cycle	开式循环

open rotor engine	开放式转子发动机
open-circuit failure	开路失效
operability margin	操作性裕度
operations room	运营室
operating ceiling	使用升限
operating condition	工作状态
operating costs per trip	每次航行的总运营成本
operating empty weight	运行空重
operating envelope	工作包线
operating instructions	操作规则
operating leasing	营业租赁
operating line	工作线
operating manual	操纵手册
operating net income	营业收入净值
operating point	工作点
operating profits	营业利润
operation across flight envelope	跨飞行包线的操作
operation centre	营运中心
operation cost guarantees	运营成本保证
operational and production readiness review	运营和生产认证审核
operational business unit	业务运营单位
operational disruption	运营中断
operational envelope	运行包线
operational excellence	运营卓越
operational interchangeability	操纵互换性
operational robustness	操作中的牢靠性
operational sovereignty	自主使用权
operational test	操纵性试验
operations engineering bulletin	运营技术通报
optimised cavities	优化的空腔
optimized wall thickness	优化壁厚
optimum SFC cycle	最佳耗油率循环
options engine order	发动机意向订单
optoelectronic device	光电子器件

orbit of shaft center	轴心轨迹
order cycle time	订购循环时间
orders placed	已下订单
orders received	收到的订单
Orenda Aerospace Company	(美国)奥伦达航宇公司
organic air vehicles	(美国计划研制的垂直起落)无人机
organic matrix composite	有机复合材料
organic paint coats	有机漆涂层
organic solvents	有机溶剂
organisation design	组织机构设计
organisational aligment	企业内部统一
Organization of Petroleum Exporting Countries	石油输出国组织
original equipment	原始设备(发动机)
original equipment manufacturer	原始设备制造商
ornithopter	扑翼机
oscillating combustion (screech，buzz)	振荡燃烧(啸叫、嗡鸣)
oscillating combustion pressure measurement	振荡燃烧压力测量
other installation items	其他安装相关内容
Otto cycles engine	奥托循环发动机
ounce	盎司
out of balance condition	不平衡条件
out of phase maintenances	不定期维护
outer air casing	外空气机匣
outer air seal	外气封
outer air swirler	外部空气涡流器
outer annulus	火焰筒外环
outer barrel	短舱外罩
outer diameter	外径
outer fan pressure ratio	外风扇压力比
outer nacelle drag	外短舱阻力
outer V blade	外 V 形叶片
outermost seal	最外侧封严
outgoing air	进出口空气
outgoing quality control	出货质量检验

outlet casing	出口机匣
outlet duct	出口导管
outlet guide vane	出口导流叶片
outlet guide vanes	出口导向器
outlet guide vanes with struts in pre-diffuser	出口导向叶片兼作扩张器的支板
outlet temperature distribution factor	出口温度分布系数
out-of-balance condition	失去平衡状态
out-of-tolerance	超差
output quality assurance	出货质量保证
output shaft	动力输出轴
output value	产值
outside air temperature	外部空气温度
outsourcing rate	外购率
over the wing engine mount	机翼上部装发动机
overall control mechanism	综合操纵机构
overall cycle parameter	整机循环参数
overall economic growth	总体经济的增长
overall efficiency	总效率
overall equipment effectiveness	设备总效能
overall performance	总体性能
overall pressure ratio	总增压比
overall test	整机试验
overboard leakage from core	核心机外泄气流
overboard leakage from fan	风扇外泄气流
overexpansion	过度膨胀
overflow valve	溢流阀
overfly manoeuvre	飞机操作
overfueling	过量供油
overhaul & maintenance	大修与维护
overhaul capacity	翻修能力
overhaul inspection	大修检查
overhaul life	翻修寿命
overhaul materials (consumable) manual	(可消耗性)大修物料手册
overhaul period	翻修周期

overhead bin	吊挂箱
overheat detection	过热探测
overheat protection	过热防护
overheat sensor	过热传感器
overheat warning light	过热报警灯
overhung moment	悬臂力矩
overload protection	过载保护
overrun	超限
overseas-listed company	境外上市公司
oversize	尺寸超大
overspeed	超转
overspeed and splitter unit	超速和隔板装置
overspeed condition	超转状态
overspeed relay	超转继电器
overspeed sensor	超转速传感器
overspeed test	超转试验
overspeed trip mechanism	超速时断油机构
overspray	过度喷涂
overtemperature	超温
overtemperature test	超温试验
over-torque	超扭矩
overwhelming preference	压倒性偏好
own supply chain	自有的供应链
oxidation	氧化
oxidation barrier coatings	抗氧化涂层
oxidation life	氧化寿命
oxidation resistance	抗氧化性
oxidation resistant coating	抗氧化涂层
oxide combustion liner	氧化燃烧室内衬
oxide dispersion strengthened	氧化物弥散强化
oxide film	氧化膜
oxides of nitrogen	氧化氮
oxides of sulphur	硫氧化物
oxidisation on titanium	钛的氧化

oxygen	氧气
ozone	臭氧
ozone test	耐臭氧试验

package B	发动机升级包 B(罗罗)
package C	发动机升级包 C(罗罗)
packaging and blanking	包装和封闭
paid-in capital	实缴资本
paint stripping solutions	漆层剥离溶剂
painting	喷涂油漆
Pampa	潘帕-战斗机(阿根廷飞机制造厂,阿根廷)
Pan American World Airways	泛美航空公司
Panavia Tornado	帕那维亚狂风战斗机
Panther	(以色列)"黑豹"无人机
paperwork certification	文字工作取证(不需做实验等硬件工作)
parallel algorithm	并行算法
parallel compressor theory	平行压气机理论
parallel genetic algorithms	并行遗传算法
parasitic losses	附加损失
part load	部分负载
part load performance	部件负载性能
part number	零件号
partial derivatives	偏导数
partial least-squares regression	偏最小二乘回归分析
partial vacuum electron beam welder	低真空电子束焊机
partially buoyant air vehicle	部分浮力航空器
particle emissions	颗粒排放
parts beyond economic repair	非经济维修组件
parts manufacturer approval	零件生产商的认证

parts manufacturing approval	零部件制造批准
parts per million	百万分率
parts service centre	零件服务中心
parts supplier	零件供应商
part-span shrouded blades	带凸肩叶片
Pascal	帕斯卡
passenger capacity	载客能力
passenger comfort	旅客舒适度
passenger emergent evacuation	旅客紧急逃生
passenger traffic flows	旅客交通流量
passengers	旅客,乘客
passive infrared system	(排气)被动红外系统
pass-off test	出厂试车
Passport	帕斯波特发动机(GE,美国)
patented innovation	专利化创新
pay-by-the-hour	按小时付费
payload	有效载荷
peak starting turbine gas temperature	最大起动涡轮燃气温度
peak to peak amplitude	双幅振幅
peak value	峰值
Pearl	珍珠发动机(罗罗德国公司,德国)
pearl river delta area	珠江三角洲地区
peel strength	剥离强度
peen forming	喷丸成型
peening and riveting	喷丸处理和铆接
peer-to-peer	终端对终端(互联网技术)
Pegasus	飞马发动机(罗罗公司,英国)
penetrant inspection	荧光检查
penetrating coloured dye	渗入着色剂
penetration	穿透深度
Pentagon's Defense Acquisition Board	(美国)五角大楼国防采购部
per annum	每年
perceived noise in decibels	感知噪声分贝
perceived noise level	可感知噪声级

percussion drilling	冲击钻孔
perfect gas	理想气体
perfect gas law	理想气体定律
perfective maintenance	完善性维护
perfluorocarbon	全氟化碳
perforate bypass duct liner	具有隔音层的外涵道
perforate hotstream liner in exhaust	排气管金属热流隔音层
perforate liner	带孔衬垫
perforate liner in bypass duct	外涵道隔音层
performance and personal development plan	业绩与个人发展计划
performance based logistics	以业绩为准的后勤保障
performance based navigation	基于飞机性能的导航
performance characteristics	性能特性
performance degradation	性能衰退
performance deterioration	性能衰减
performance improvement	性能改善
performance indicator	性能指标
performance lapse	执行失误
performance lapse with speed	性能随速度衰减
performance metrics	业绩指标
performance review	业绩审核
performance review institute	绩效考核机构
performance seeking control	寻找并控制最佳性能状态
performance trend analysis	性能趋势分析
performance-based organization	以业绩为准的机构
performance-oriented workplace	以业绩为准的工作
periodic table	元素周期表
Perm Design Bureau	(俄罗斯)彼尔姆发动机设计局(现为彼尔姆发动机公司)
permanent magnet alternator (engine dedicated generator)	永磁交流发电机(发动机专用发电机)
permanent magnet brushless motor	永磁无刷电机
permanent magnet motor	永磁电机
permanent magnetic alternator	永磁交流电机

perpetual growth	持续增长
persistent，bio-accumulative and toxic	持久性、生物蓄积性和毒性
perspex sector	有机玻璃扇形段
petrol engine	汽油发电机
Phantom	鬼怪-战斗机(麦道公司,美国)
phase correction accuracy	相位校正精度
phase displacement meter	相位扭矩计
phase Doppler analyzer	相位多普勒分析仪
Phenom	飞鸿-公务机(安博维公司,巴西)
phonic wheel	音轮
photoelasticimetry	光弹测量学
physical and functional interchangeabilities	物理和功能互换性
physical interface control document	物理界面控制文件
physical vapor deposition	电子束物理气相沉积
Piaggio Aero Industries	(意大利)比亚乔航空工业公司
Picador	(以色列)"斗牛士"旋翼无人机
pick & pack	验货与包装
pickling	酸洗
Pilatus	皮拉图斯-公务机(皮拉图斯飞机公司,瑞士)
pilot burner	辅燃烧室
pilot fuel	引燃油
pilot mixing inner port	辅混区内端口
pilot parts production	部件的试生产
pilot spraybar	先导喷油管
pilotless drone	无人机
pilot's control lever	驾驶员操纵杆
pins-slotted	圆柱销槽
pipeline	油气管线
pipeline control solenoid valve	油气管线控制电磁阀
pipeline strain test system	油气管线应变测试系统
pipes and harness	管线
piston	活塞
piston cooling oil pump	活塞冷却油泵

piston engine	活塞发动机
piston engine cooling	活塞发动机冷却
piston ring	活塞环
piston ring seal	金属涨圈密封
piston-type pump	活塞式泵
pitot intake	压差计进气口
pitot-static system	皮托静压系统
pitot-tube anti-icing	空速管防冰
pitot-type intake	空速管型进气道
pitting	气蚀,凹痕
pivot bearing	枢轴承
pivot thrust reverser	转门式反推装置
plain nozzle	普通喷管
plain orifice atomizer	直射喷雾器
plain-orifice nozzle	直喷喷嘴
plan，do，study，act	计划、执行、检查、调整
plane angle	平面角
plane cascade test	平面叶栅试验
planetary geared drum	行星齿轮传动滚筒
planned hours	计划时数
plant supplier quality engineer	工厂供应商质量工程师
plasma arc surfacing	等离子弧堆焊
plasma arc welding	等离子弧焊接
plasma engine	等离子发动机
plasma measuring	等离子测量
plasma propel	等离子推力器
plasma propulsion	等离子推进
plasma sprayed	离子喷涂
plasma sprayed ceramic coatings	等离子喷涂陶瓷涂层
plasma sprayed coating	等离子喷涂涂层
plasma spraying equipment	等离子喷涂设备
plasma torch	等离子喷枪
plasma welding	等离子焊接
plastic deformation	塑性变形

plateau airport	高原机场
platform film cooling holes	叶根平台气膜冷却孔
platinum group metals	铂族金属
pleated wire mesh filter	折叠丝网油滤
plenum chamber	稳压室
plenum chamber burning	增压室燃烧
plug nozzle	塞式喷管
plug-in board	插卡结构
plunger pump	柱塞泵
plunger-type pump	柱塞式泵
ply cutting technology	板材下料技术
pneumatic average pressure	气动平均压力
pneumatic brake system	空气刹车系统
pneumatic control	气动控制
pneumatic cylinder	汽缸
pneumatic flow field	气动流场
pneumatic inputs	气动输入
pneumatic power	气动功率
pneumatic torque multiplier	气动扭矩放大器
Poisson's ratio	泊松比
poka yoke principle	防错原则
polished features	抛光特性
political, environmental, social and technological influence	政治,环境,社会和技术的影响
polyamide composite	聚酰胺基复合材料
polycarbonate composite	聚碳酸酯复合材料
polycrystal equiax	等轴多晶体
polyester infill panels	聚酯充填板
polygonal casing	多边形机匣
polyimide composite	聚酰亚胺复合材料
polymer ceramic composite coatings	聚合物与陶瓷复合材料涂层
polymer matrix composite	树脂基复合材料
polymerization of monomer reactants	单体反应物聚合
polytropic efficiency	多变热效率

poor reliability	可靠性差
pop-up menu	弹出式菜单
porosity	气孔
port	左舷
port engine	左侧发动机
portable video system	便携式视频系统
position accuracy	定位精度
position resolution	位置分辨率
positive curved blade	正弯曲叶片
positive incidence stall	正迎角失速
post certification testing	取证后的试验
post-certification activity	合格审定后的活动
post-certification engineering	取证后工程
potentially longer life	长寿命潜力
potentiometer	电位计
pound	磅
pound per square inch	每平方英寸磅力
pounds force	磅力
pounds-foot（feet）	磅力英尺
pounds-inch（es）	磅力英寸
pouring tube passage	浇注管通道
powder disc	粉末冶金轮盘
PowerJet	（法国斯奈克玛公司与俄罗斯土星公司的合资）喷气动力公司
powder metal	粉末金属
powder metallurgic material	粉末冶金材料
powder metallurgy	粉末冶金
powder metallurgy disc	粉末冶金盘
power	功率
power booster	助力器
power by hydraulic	液压动力
power by the hour	（罗罗船舶）动力系统计时付费包修合同
power by wire	功率电传

power density	动力密度
power drive unit	电源驱动装置
power electronic module	功率电子模块
power indicating sensor	功率显示传感器
power level	功率水平
power lever angle	功率杆角度
power limiter	功率限制器
power management control	动力管理控制
power management system	动力管理系统
Power Mill	(英国 DELCAM 公司的)三维数控加工软件
power off	关机、断电
power optimised aircraft	动力优化飞机
power output shaft	动力输出轴
power plant station	发电站
power plant system	动力装置系统
power setting	功率设置
power supply unit	动力供应装置
power take off	功率输出
power take off device	功率提取装置
power take off extreme	极端功率输出
power to weight ratio	功重比
power turbine	动力涡轮
power turbine inlet temperature	动力涡轮进口温度
Power8	空客 2007 年重组计划的名称
power-by-the-hour airline operation	发动机具有计时付费包修合同的航线运营
power-by-wire	电传动力
PowerJet	喷气动力公司(斯奈克玛与土星合资公司,法国/俄罗斯)
power-off-immersed	关油门状态
power-off-retracted	撤销关油门指令
powerplant	动力装置
powerplant assembly	动力装置的组装

powerplant integration	动力装置一体化
practical fixing	可行的解决方案
Pratt & Whitney Canada Corporation	普惠加拿大公司（加普惠，加拿大）
Pratt and Whitney	（美国）普惠发动机公司
precession	进动
precising casting	精铸
precising casting facility	精铸车间
precision forging	精锻
precision lathe	精密车床
pre-compression	预压缩
pre-cooler	预冷器
pre-cooler exhaust	预冷器热气排放
precooling of cooling air	冷却空气预冷
predesign tool	预设计工具
predictable budget cycle	可估成本维护过程
predictable costs	可预估的修理费用
predictable failure	可预测失效
predicted cyclic life	预估循环寿命
predicted growth	预测的增长
predicted safe cyclic life	预估安全循环寿命
predictive maintenance	预知维护
pre-eminent budget airline	优秀的低成本航空公司
pre-fabricated module	预制模块
pre-fabricated parts	预制零件
preferential trade agreement	优惠贸易协定
preferential trade area	优惠贸易区
preferred bidder status	优先投标者身份
preferred stock equity	优先股权益
preferred supplier	优选供应商
pre-first flight risk assessment	首飞之前风险评估
preheater	预热器
prelaunch	预启动
preliminary concept review	初级方案评审
preliminary design	初步设计

preliminary flight rating test	首次飞行状态试验
preliminary material review	初级物料审核
preliminary type board meeting	型号委员会预备会议
Premier	首相-公务机(塞斯纳飞机公司,美国)
premium quality products	优质的产品质量
premix	预混
premix burner	预混火焰筒
premix chamber	预混腔
premixed flame	预混火焰
premixed prevaporized combustor	预混-预蒸发式燃烧室
prepaid expenses	待摊费用
prerotating	(旋翼)预旋
Presidency of Civil Aviation	(沙特阿拉伯王国)民航局
pressure	压力
pressure at engine intake	发动机进气口压力
pressure balance pipes	平衡压力输气管
pressure belt	压力带
pressure calibrator	压力校准仪
pressure controller	压力调节器
pressure differential transmitter	压差变送器
pressure drop	压力损失
pressure drop control spill valve	压降控制溢流活门
pressure drop indicator	压差指示器
pressure drop relay	压差继电器
pressure energy added	增加压力能
pressure filter	高压油滤
pressure fueling station	压力加油舱口
pressure gradient	压力梯度
pressure gauge	压力计
pressure jump	压力跃变
pressure loss	压力损失
pressure loss coefficient	压力损失系数
pressure loss factor	压力损失因素

pressure measurement in distortion flow field	畸变流场的压力测量
pressure quenching machines	压力淬火机
pressure raising shut-off valve	增压截流阀
pressure rake	测压排管
pressure ratio	压力比
pressure ratio based engine control	利用压比控制发动机
pressure ratio growth potential	压比增长潜力
pressure ratio regulator	压比调节器
pressure recovery	压力恢复
pressure reducer	减压器
pressure regulating shut-off valve	压力控制截流阀
pressure regulating valve	压力控制阀
pressure regulator	压力调节器
pressure relief door	减压门
pressure relief valve	减压阀
pressure relief valve oil system	滑油系统减压阀
pressure sensor rake	压力传感器排管
pressure sensors	压力传感器
pressure side	压力面
pressure thrust	压力推力
pressure vessel integrity	压力容器完整性
pressure volume diagram	压力-体积图
pressure wave	压力波
pressure-filled supply tank	充压滑油箱
pressure-jet injectors	高压喷嘴
pressurized turboprop	加压涡桨发动机
pressurized water reactor	压水反应堆
pressurizing valve	高空活门
pre-stress	预应力
pre-swirl nozzle	预旋喷嘴
pre-system development and demonstration programme	先期系统开发和验证计划
preventative maintenance plan	预防性维护计划
preventive maintenance	预防性维护

primary air	主流空气
primary air scoop	主空气戽斗
primary air swirler	主离心式空气喷嘴
primary combustion zone	主燃区
primary design review	初级设计审核
primary flight control system	飞行控制主系统
primary hole	主燃孔
primary solidification structure	一次结晶组织
primary structure	主承力结构
primary zone	主燃区
prime contractor	主承包商
printed circuit board	印刷电路板
prior counteraction	预先反作用
private financing initiative	私人融资计划
pro-active approach	预测方案
proactive maintenance	主动维护
probe access panel	探头入口
probe microphone	探管传声器
problem change report	问题变化报告
problem presentation sheet	问题说明表
procedure instruction	流程说明
process/product control document	工艺/产品控制文件
process change notice	工序改动通知
process council	流程委员会
process critical parameters	关键过程参数
process data sheet	流程数据表
process excellence	卓越流程
process failure modes and effects analysis	故障模式和影响分析流程
process improvement request	流程改善请求
process specification	工艺规范
procurement information booklet	采购信息手册
procurement policy document	采购策略文件
product/project change authority	产品/项目变更审批部
product change board	产品变更委员会

product conformance improvement team	产品一致性改善组
product conformity process adherence	产品一致性工艺规定
product control authority	产品控制部门
product data management	产品数据管理
product definition document	产品定义文件
product focus	产品重点
product innovation and lifecycle management	产品开发与全寿命周期管理（罗罗）
product integrity committee	产品完整性委员会
product introduction management	产品推介管理
product introduction management gateway review	产品推介管理过门审核
product introduction process	产品推介流程
product life-cycle management	产品全寿命周期管理
product lifecycle management best practice	产品全生命期管理最优方法
product lifecycle management operating procedure	产品全生命期管理运作程序
product performance metrics	产品性能优化
product portfolio	产品范围
product quality deficiency report	产品质量缺陷报告
product requirements document	产品规范
product strategy review	产品战略审核
product technical controller	产品工艺管理员
product test specification	产品检测规范
product warranty	产品担保
production & material control	生产和物料控制
production acceptance test limits	产品合格试验极限
production batches	生产批次
production engine	产品型发动机
production engine test	产品型发动机试车
production equipment for liquid propellant	液体推进剂生产设备
production line bottleneck	生产线瓶颈
production management	生产管理
production organisation approval （EASA Part 21）	（欧洲航空安全局第 21 部）生产机构审批

production pipeline inventory	生产线存货
production plan control	生产计划控制
production ramp-up rate	生产速度
production readiness guidelines	生产准备指南
productive assets	生产性资产
productive verification	产品验证
professional excellence	杰出专业人才
profile factor	截面系数
profile rings	型面环
profile/airfoil	叶型
profit before interest and tax	息税前盈利
profit before tax	税前纯利
profitability	收益率
prognostics and health management unit	诊断与健康管理单元
prognostics and health monitoring	诊断与健康监测
program execution	项目执行
programme agreement	项目合作协议
programme levy	项目管理费
programme on schedule	按计划开发
project champion	项目冠军,最优质量项目
project chief design engineer	项目总设计师
project chief engineer	项目总工程师
project cost estimate	项目成本预估
project infrastructure	项目组织结构
project interactive analysis and optimization	(初级飞机设计程序)互动式项目分析和优化
project kick-off meeting	项目启动会议
project on government oversight	(美国)政府监督项目
project policy document	项目政策文件
project requirements document	项目规范
project start	项目启动
project status	项目现状
project sub-system requirement	项目子系统要求
project team leader	项目组长

project zero	（罗罗公司提高运营遣派率的项目）零中断率项目
proof of concept	方案验证
proof strength	屈服强度
prop flange	支撑法兰盘
prop shaft	驱动轴
propagation of cracks	裂纹扩展
propellant filling system	推进剂加注系统
propeller	螺旋桨
propeller area ratio	螺旋桨面积比
propeller back	螺旋桨背部
propeller brake	螺旋桨刹车装置
propeller characteristic operating status	螺旋桨特征工作状态
propeller characteristics	螺旋桨特性
propeller control system	螺旋桨控制系统
propeller control unit	螺旋桨控制器
propeller controller unit	螺旋桨控制装置
propeller feathering	顺桨
propeller geometric parameters	螺旋桨几何参数
propeller handling tools	螺旋桨装卸工具
propeller loading test	螺旋桨负荷试验
propeller motion parameters	螺旋桨运动参数
propeller nut	螺旋桨锁帽
propeller parking brake	螺旋桨止动装置
propeller pitch fixing	螺旋桨定距
propeller pitch variation	螺旋桨变距
propeller pressure side	螺旋桨压力面
propeller research tunnel	研究螺旋桨的风洞
propeller reversing	反桨
propeller shaft revolution indicator	螺旋桨轴转速计
propeller speed governer	螺旋桨调速器
propeller speed limiter	螺旋桨限速器
propeller suction side	螺旋桨吸力面
propeller thrust	螺旋桨推力

propelling nozzle	推进喷管
propfan	桨扇
propfan engine	桨扇发动机
propfan test assessment	桨扇发动机测试评估
proposed master programme	开发计划建议
propulsion system	推进系统
propulsion system safety board	推进系统安全理事会
propulsion systems laboratory	推进系统实验室
propulsion/AC integration	推进系统机身一体化
propulsion-controlled aircraft	(美国国家航空航天局)用推进系统控制的飞机
propulsive efficiency	推进效率
propulsor engine	喷气推进发动机
proton exchange membrane	质子交换膜
proton exchange membrane fuel cell	质子交换膜燃料电池
prototype	原型机
prototype engine	原型发动机
prototype rolled out on schedule	原型机按时出库
prove structural integrity	验证结构的完整性
proven technology	经过检验的技术
pseudo loss coefficient	虚拟损失系数
pseudo random spectrum	伪随机谱
psi absolute	绝对压力(磅每平方英寸)
psi differential	压差(磅每平方英寸)
psi gauge	传感器测量压力(磅每平方英寸)
public offer	公开报价
public ownership	上市公司
pullers	拆卸器
pulse assembly line	脉冲总装线
pulse detonation engine	脉冲爆震发动机
pulse detonation engine combustion	脉冲爆震发动机燃烧
pulse duration	脉冲持续时间
pulse jet engine	脉冲喷气发动机
pulse line	脉动式生产线(波音)

pulse repetition frequency	脉冲重复频率
pulse solid rocket motor	脉冲固体火箭发动机
pulse welding	脉冲焊接
pulsed arc exciter	脉冲激弧器
pulsed electron beam welding	脉冲电子束焊
pulsed Nd	脉冲钕
pulsed wave	脉冲波
pump recharging	油泵再次充油
punctuality	正点率
punitive damages	惩罚性的损失赔偿
purchase by inload	从自己的供应商采购
purchase order	订单
purchasing order	采购订单
purchasing policy & practice	采购方针和方法
Pure Power	洁净动力（普惠）
purging	吹除
pushers	顶推器
Pyestock altitude test facility envelope	位于派斯多克的高空台包络线
pylon	标塔（火焰稳定器）
pylon fairing	吊架整流罩
pylon service	吊架接口偶合设备
pylon structure	吊架结构
pyrolytic deposition	热解沉积
pyrometer	高温计

QC2 departure noise rules	QC2 起飞噪声规则
quadrature amplitude modulation	正交调幅
qualification test	定型试验
quality acceptance standard	质量验收标准
quality acceptance standard non-destructive testing	无损检测质量验收标准

quality and affordability	质量和可承受性
quality assessment criteria	质量评估标准
quality assurance plan	质量保证计划
quality assurance procedure instruction	质保流程条例
quality assurance representative	质量保证代表
quality auditing	质量审核
quality control	质检
quality control test procedure	试验的质量控制流程
quality & delivery performance	质量和交付表现
quality escape	质量逃逸
quality expectation	质量预期值
quality management system	质量管理系统
quality notification	质量通报
quality operating procedure	质量管理程序
quality plan	质量计划
quality renaissance leader	质量复兴项目主管
quality resolution co-ordinator	质量解决协调人
quality standard	质量标准
quality systems internal auditor course	质量体系内审员培训课程
quality，cost，delivery，responsiveness	质量、成本、交货、反应
quality/reliability/service	质量/可靠度/服务
quality-assurance capability	质量保证能力
quantitative maintainability	定量维修性
quantity	数量
quartz-fibre-reinforced composite	石英纤维增强的复合材料
quality assurance division	质量保证处
quench	淬水
quench air	风冷淬火
quench PA	水溶性聚合物淬火剂
quench zone mixing	贫油区油气混合
quick cut-off valve	快速截止阀
quick engine change	快速更换发动机
quick engine change unit	快速发动机更换装置
quick release	快速发布

quick response code	（日本研发的）快速响应矩阵条码
quick-drying cleaning fluid	快干清洗液
quiet airplane technology	静音飞机技术
quiet clean general aviation turbofan	静音低排放通航发动机
quiet clean short-haul experimental engine	静音低排放短途实验发动机
quiet climb system	静音爬升系统
quiet engine program	静音发动机计划
quiet experimental short takeoff and landing	静音实验性短距起降
quiet fan	静音风扇
quiet high speed fan	静音高速风扇
quiet short-haul research aircraft	静音短距研究机
quiet short take-off and landing	静音短距起降
quiet supersonic transport	（美国 Aerion 集团）静音超音速客机
quiet technology demonstrator	静音技术验证机（波音）
quiet technology demonstrator 2	第二号静音技术验证机（波音）
quintuple pass multi-fee internal cooling	五通道复式内冷
quota count	限额数

racing car	赛车
radar absorbing material	吸收雷达波材料
radar chart	雷达图（多轴平面图）
radar data processor	雷达数据处理器
radar scattering	雷达散射
radar-evading stealth fighter	逃避雷达的隐形战斗机
radial	径向
radial deflection	径向位移
radial drive	径向传动
radial driving shaft	径向传动轴
radial dual swirler	径向双旋流器
radial inflow turbine	向心式涡轮
radial rear stage	径向后置级

radial sidewind cooling of disk	轮盘径向吹风冷却
radial temperature distribution factor	径向温度分布数
radial total pressure distortion sensitivity	径向总压畸变灵敏度
radial turbine	径流式涡轮
radially disposed vanes	径向配置的导向叶片
radiation barrier coating	防热辐射涂层
radiation hardened	抗辐射加固的
radiation pyrometers	辐射高温计
radiation-hardened electronic devices and components	抗辐射加固的电子装置和部件
radio altimeter	无线电高度表
radio frequency identification	射频识别技术
radio frequency identification tags	射频识别标签
radio interference	无线电干扰
radio-active cleaning material	放射性清理材料
Rafale	阵风-战斗机(达索公司,法国)
rain ingestion	吞雨
rake	压力探头安装架
raked winglet	后掠式小翼
Raleigh flow	瑞利流
ram air	冲压空气
ram air turbine	冲压空气涡轮
ram jet engine	冲压式喷气发动机
ram pressure	冲击压力
ram turbine	冲压涡轮
Raman spectroscopy	拉曼光谱
ram-compressed rotor	冲压转子
ramjet engine	冲压喷气发动机
ramp boundary layer bleed system	斜板附面层吹除系统
ramp weight	最大滑行重量
ram-top fixture	升降夹具
random force	随机力
random noise	随机噪声
range factor	航程因子

range temperature control unit	温度范围控制器
Rankine cycle	兰金循环
rapid descaling system	快速去除封严系统
rapid deterioration	迅速衰退
rapid prototyping	快速原型制造
rapid reducing valve	急降活门
rapid repeat start	快速重复起动
rapid response line repairs and module "hospital shops"	快速反应的航线修理和单元体维修车间
rapid tooling	快速模具
rapidly solidified Ti-base brazing filler	快速凝固钛基纤料
Raptor	(美国)F-22猛禽战斗机
ratchet adaptor	棘齿转换器
rate of climb indicator	升降速率表
rate of diffusion	扩压度
rate per flight hour	按小时付费
rate schedule	费率表
rated horsepower	额定马力
rated output capacity	额定输出
rated thrust	额定推力
rating structure	推力设置结构
rationalised process specification	优化的流程规范
rationalised quality standard-component specific	特殊部件质量标准-组件特定性(罗罗)
rationalised quality standard-component type	部件型号质量标准-组件型号(罗罗)
rationalised quality standard-generic	通用质量标准(罗罗)
rationalised quality standard-process	工艺质量标准(罗罗)
rationalised quality standards	质量标准(罗罗)
raw materials	原材料
Raytheon Company	(美国)雷神公司
reach implementation project	(欧盟化学品注册评估和审批政策)实施项目
reacting flow	回流
reaction	反作用

reaction control system	反作用控制系统
reaction control valve	反应控制活门
reaction turbine	反动式涡轮
reaction-bonded silicon nitride	反应烧结氮化硅
reactor core materials	反应堆芯材料
read-across	交叉参照
real time acoustics	实时声学分析
real time alarming	实时报警
real time communication	实时交流
real time modelling	实时模型
real time simulation model	实时模拟模型
real time troubleshooting	实时故障排除
real time visibility	实时显示
rear bearing chamber	后部轴承腔
rear bearing support structure	后轴承支承结构
rear casing	后部机匣
rear fuselage installation design	机身后部安装设计
rear half casings	机匣下半部
rear journal bearing	后部滚棒轴承
rear mount	后安装节
rear mount and LPT casing	后安装节和低压涡轮机匣
rear mount pylon interface	后安装节界面
rear mounted outer swirler	安装在后部外侧的涡流器
rear mounting	尾部安装
rear mounting beam	后部安装梁
rear mounting bracket	后安装节托架
rear outer interface	后部外接口
rear stub shaft	后短轴
rear support ring	后支撑环
rear-facing step	后向台阶(火焰稳定器)
rearward	向后
reattachment	再附着
recast layer	重铸层
receipt inspection	收货验货

reciprocating engine	往复式发动机
reciprocating piston engine	往复式活塞发动机
reciprocating pump	往复泵
recirculation zone	回流区
recirculatory oil system	循环滑油系统
recommended spare parts list	首批推荐备件清单
recovery from surge	退喘
recrystallizing heat treatment	再结晶热处理
rectifier housing	整流器罩
recuperated	回热
recuperator	回热器
RecurDYN	(韩国)多体动力学分析软件
recurring cost	经常性成本
redline speed	极警速度
redline temperature	极警温度
reduce to produce	降耗减重生产
reduced acoustic mode scattering engine system duct	低噪声散射发动机管道系统
reduced block time	缩短的轮挡时间
reduced overall length	总长度较小
reduced shock losses	降低激波损失
reduced supplier defect	减少供应商提供的不合格品
reduced temperature configuration	降温构型叶片
reduction gear assembly	减速齿轮组件
reduction gear box	减速齿轮箱
reduction gearbox with two propeller shafts	双桨轴减速器
redundancy managing	余度管理
re-fanned variant	换风扇型
reference pressure relief valve	参考压力减压活门
refrigerant injection	冷却液喷射
refurbishment program	更新计划
regenerative cooling	再生冷却
regenerator	回热器
regional aircraft	支线飞机

regional jet	支线喷气式飞机
regional regulatory authority（Rolls-Royce）	区域性管理机构（罗罗）
regional traffic	支线交通
registration，evaluation and authorisation of chemicals	（欧盟）化学品的注册评估和审批草案
regular process surveillance	定期进行的程序审核
regulation classification	条例分类
reheat	加力
reheat control	加力控制
reheat fuel control unit	加力燃油控制器
reheater	加力燃烧室
Reid vapour pressure	雷德蒸发压力
reinforced carbon-carbon	增强碳碳复合材料
reinforced plastic	增强塑料
reinforcement of implementation	强制实施
relative density	相对密度
release note	发货通知、发布说明
release note（certification of completion）	产品发放注释单（完工证明书）
release of the propeller	螺旋桨飞出
reliability	可靠性
reliability and safety documentation	可靠性和安全性文件
reliability growth	可靠性增长
reliability maturity	可靠性的成熟
relief valve	泄压阀
relight	再点火
relight capability	再点火能力
relighting in flight	空中再点火
remaining surge margin	剩余喘振裕度
rematch	重新匹配
remote lift fan	远距升力风扇
remote augmented lift system	远距增升系统
remote control unit	遥控装置
remote data concentrator	远程数据采集器
remote fault diagnosis	远程故障诊断

remote monitoring and diagnostic system	远程监控和诊断系统
remote testing	遥控测试
remotely operated vehicle	遥控车
remotely piloted vehicle	遥控飞行器
remove/replace time	拆换时间
renewable	可再生能源
renewal by repair	维修更新
repair & overhaul service	维修和大修服务
repair and overhaul	修理和大修
repair depth	维修深度
repair of spares	备件修理
repair technologies	修理技术
replacement aircraft study	后继机研究
repositioned	重新布置
request for alteration	变更申请
request for comment	请求注释
request for engineering source approval	要求批准工程资源
request for information	信息征询书
request for offer	要求报价
request for proposal	招标
request for qualifications	质量要求
request for quotation	询价
request for variation	变更申请
required inspection item(s)	要求检验的项目
required net thrust per engine	每台发动机所需净推力
resaturated pyrolized carbon-carbon materials	多次浸渍的热解碳/碳材料
research & development	研究与开发
research and innovative technology administration	(美国交通运输部)研究和技术创新局
research and technology	研究和技术
research，development and manufacturing	研发和生产
reserve factor	安全系数
reserve fuel	备份油量
reserve tank	副油箱

reservoir capacitor	储能电容器
reset cam	调整凸轮
residence time	停留时间
residual fuel	剩余油量
residual life	剩余寿命
residual life on wing	在翼剩余寿命
residual strain	残余应变
residual stress	残余应力
resin film infusion	树脂膜熔渗成形
resin impregnated with fibre	树脂浸渍纤维
resin matrix	树脂基体
resin penetration model	树脂渗透模型
resin prepreg	树脂预浸料
resin transfer molding	树脂传递成型
resin transfer molding composite fan	树脂传递成型复合风扇
resistance bulb thermometer	电阻温度传感器
resistance temperature device	电阻测温装置
resistance temperature sensor	电阻测温器
resistance to failure	防失效
resistance to galling	防擦伤
resolve customer problem	解决客户问题
resonance	共振
resonance damping	共振阻尼
responding curve	反馈曲线
responses to technical queries	解答技术问题
rest of the world	世界其他地方
restart	再起动
retained earnings	留存赢利
retained or not retained	保留或非保留
retaining pin clearances	(叶片)保持销的间隙
retaining pins	固定销钉
retaining setscrew	定位保持螺钉
retractable landing gear	可收放起落架
retrofit control	控制改造

retrofitable package	翻新包
return on assets	资产收益
return on capital	资本收益率
return on capital	资本利润率
return on capital employed	已动用资本回报率
returned material approval	退货验收
return-to-tank valve	回油箱阀
revamped plane	改进的飞机
revenue passenger kilometer	每旅客每公里收费
revenue per available seat mile	每座英里收入
revenue sharing partnership	收益分成伙伴关系
revenue-earning potential	赢利潜力
reverse engineering	逆向工程
reverse flow combustion system	回流式燃烧系统
reverse flow combustor	回流燃烧室
reverse flow system	回流系统
reverse thrust	反推力
reverse thrust load	反推载荷
reverse thrust select lever	反推力选择手柄
reversible pitch propeller	可变螺距螺旋桨
revert material	返回料
revolutionary approach to time critical long range strike	（美国）远程打击时间敏感目标的导弹项目
revolutionary concepts in aeronautics record group	航空记录组的革命性方案
revolutionary turbine accelerator	革命性涡轮加速器
revolutions per minute	每分钟转速
revolving credit	周转性信贷
Reynolds effects	雷诺效应
Reynolds number index	雷诺数指数
Reynolds number	雷诺数
Reynolds number ratio	雷诺数比
rich zone	富油区
rich-quench-lean combustion	富油—猝熄—贫油燃烧

rig test	部件试验
right first time	一次成功
right hand	右手
right hand side	右手侧
rigid borescope	刚性内视镜
rigid rotor	刚性转子
rigid technical endoscopes	刚性工程内视镜
rim pull	边缘拉力
rim segment	轮缘扇形块
ring engine-mount	发动机安装环
ring forgings	环锻件
ring rolled	轧制环
ring rolls	环轧机
rings	环件
riser patterns	冒口型模
risk and revenue sharing partner	风险和收益分成伙伴
risk and revenue sharing partnership	风险和收益分成伙伴关系
risk assessment procedure	风险评估方法
risk based inspection	风险检查
risk exposure	出险
risk management	风险管理
risk mitigation	降低风险
risk priority number	风险顺序数
risk revenue sharing	风险收益共担
risk sharing partner	风险共享伙伴
riveting pressure	铆接压力
riveting	铆接
robust material	经久耐用的材料
rocket assisted take-off	火箭辅助起飞
rocket based combined cycle	火箭基组合循环
rocket engine	火箭发动机
rocket motor	火箭发动机
rocket-type combustion chamber	火箭型燃烧室

Rockwell C hardness	（罗克韦尔公司定义的）洛氏 C 级硬度
Rockwell hardness	洛氏硬度
roll posts	转动姿态
rolled ring	轧环形件
roller bearing	滚子轴承
rolling circle arc	滚动圆弧
rolling mill	滚轧机
roll-out	（飞机）出库
rollout phase	实施阶段
roll-post aerodynamics design	转动姿态气动设计
Rolls by-pass	罗尔斯旁路
Rolls E. L. Turbofans Ltd.	（罗罗子公司负责发动机支持和服务）罗尔斯涡扇发动机公司
Rolls Royce North America Incorporated	（美国）罗尔斯·罗伊斯美国公司
Rolls-Royce（UK Aerospace）	英国罗尔斯·罗伊斯公司
Rolls-Royce Brazil	罗罗巴西公司
Rolls-Royce Canada Repair & Overhaul	罗罗加拿大修理和大修公司
Rolls-Royce Corporation	罗罗美国公司
Rolls-Royce Deutschland Ltd & Co. KG	罗罗德国有限公司
Rolls-Royce Energy Customer Service Business	罗罗能源客户服务业务
Rolls-Royce Energy Gas Turbines Canada	罗罗能源加拿大燃机公司
Rolls-Royce Energy Package & Power Systems	罗罗能源系统集成和动力系统公司
Rolls-Royce Energy Systems	罗罗能源系统公司
Rolls-Royce Engine Services Oakland	（美国）罗罗奥克兰能源服务公司
Rolls-Royce engineering specification	罗罗技术规范
Rolls-Royce Liberty Works	罗罗北美创新中心
Rolls-Royce managed service provider	罗罗公司管理的服务商
Rolls-Roycc marinc	罗罗公司舰船业务
Rolls-Royce marine electrical systems	罗罗公司舰船电气系统
Rolls-Royce marine power	罗罗公司舰船动力
Rolls-Royce marine power operations Ltd.	罗罗公司舰船业务公司
Rolls-Royce marine systems	罗罗公司舰船系统
Rolls-Royce Power Engineering Ltd.	罗罗公司动力工程公司

Rolls-Royce production system	罗罗公司生产系统
Rolls-Royce Public Limited Company	(罗尔斯·罗伊斯股份公司)罗罗公司
Rolls-Royce quality management system	罗罗公司质量管理系统
Rolls-Royce Turbomeca	罗罗透博梅卡合资公司
root cause analysis	根本原因分析
root cause report	根本原因报告
root pressure ratio	根部增压比
rootcause analyse	根源分析
Rosin-Rammler model	罗辛-拉姆勒模型(粒子尺寸分布)
Rostec State Corporation	(俄罗斯)技术国家集团
rotary atomizer	甩油盘
rotary drum regenerator	转鼓式回热器
rotary engine	转缸式发动机
rotary variable differential transformer	旋转可调差分变压器
rotary wing	旋翼
rotating air sealing	旋转式空气密封
rotating and stationary part	转动和静止部件
rotating cascade test	旋转叶栅试验
rotating casing	旋转机匣
rotating drum	旋转鼓筒
rotating guide vane	旋转的导向叶片
rotating parts cycle life	转动部件循环寿命
rotating ramjet	旋转冲压发动机
rotating rear	后旋转件
rotating stall	旋转失速
rotational speed	旋转速度
rotative forgings	转动部件的锻件
rotor	转子
rotor 1 angled fin	第一级转子叶片篦齿加前倾角度
rotor balancing	转子平衡
rotor blade deicing	螺旋桨叶片除冰
rotor bowl	(发动机转子停车后)转子热变形挠曲
rotor brake	旋翼制动

rotor dovetail leakage	转子楔形渗漏
rotor failure	转子失效
rotor hub	转子轮毂
rotor inlet temperature	转自进口温度
rotor locking test	转子锁定试验
rotor self-centralisation	转子自动定心平衡
rotor shaft	旋翼轴,转子轴
rotor spacer	转子隔圈
rotor speed sensor	转子转速传感器
rotor structure	转子结构
rotor support scheme	转子支承设计图
rotor support structure	轴子支承结构
rotor track and balance	转子跟踪和平衡
rotor vehicle	旋翼飞行器
rotordynamic analysis	转子动力学分析
rotor-stator interaction	转子静子相互作用
rotor-stator interaction noise	转子静子干涉噪声
rough machined rings	粗加工环
rough order of magnitude	粗略估计
round head rivets	半圆头铆钉
route to process excellence	卓越流程项目指南(罗罗)
route-proving flight	试飞新航线
routine maintenance	例行维护
Royal Australian Air Force	(澳大利亚)皇家空军
rub in and sealing coating	耐磨封严涂层
rub indictor	磨蹭指示器
rubbing and contact effect	摩擦和接触的影响
run chart	进程图、运行图
run down，rundown	降转速
rundown resonance	降转速通过共振点
running down time of engine rotor	发动机转子减速停车所需时间
running in	磨合
running start	起动
runout	偏离

runtime system	(软件)运行系统
runway independent aircraft	超短距起降飞机
runway visual range	跑道能见度
rupture of casing	机匣破裂
rupture of engine mount	发动机安装节破裂
Ruslan	鲁斯兰-运输机(安东诺夫公司,乌克兰)
Russia regional jet	俄国支线喷气机
Russian acronym: ecology and progress	俄语缩写:生态与进步
RWTH Aachen Institute of Jet Propulsion and Turbomachinery	(德国)莱茵-威斯特法伦亚琛工业大学喷气推进和叶轮机械研究所

S1 and S2 relative stream surfaces	S1 与 S2 相对流面
safe valve	安全活门
safety life design	安全寿命设计
safety management system	安全管理系统
safety margin	安全裕度
Safety Review Board	安全审查委员会
safety valve	安全阀
safety-critical components	关键性安全部件
SAFRAN	法国赛峰集团
sales & operations planning	销售和运营计划
sales and operations review board	(罗罗公司)销售和运营审查委员会
sales discount & allowance	销售折扣与折让
sales volume	销售额
Salyut Machine Building Association (Salyut)	(俄罗斯)礼炮航空发动机联合体
sample debugging	小样调试
sample regression coefficient	样本回归系数
sample size	样本大小
sample standard deviation	样本标准差
Samsung Aerospace Industries Ltd.	(韩国)三星航宇工业公司

sand cast casings	砂模铸造机匣
sand cores	砂芯
sand filter device	防沙装置
sand mould	砂模
sapphire fibre-optic structure	蓝宝石光纤结构
saturated steam	饱和蒸汽
saturation pressure	饱和压力
Sauter mean diameter	索太尔平均直径
scaled development of an engine	发动机按比例放缩衍生
scaling	缩放
scaling parameter group	缩放参数组
scanning data acquisition module	扫描式数据采集模块
scanning electron microscopy	扫描电子显微镜
scanning tunneling microscope	扫描隧道显微镜
scarfed intake	下唇凸出的进气道
scatter diagram	散布图
scavenge	回油
scavenge filter	回油滤
scavenge oil pump	废油泵
scavenge pump	回油泵
scavenge screens	回油滤网
scavenge system	回油系统
scavenging pump	驱气泵
scheduled airlines	定期航班
scheduled inspection	定期检查
scheduled maintenance	定期维护
schematic plot	简图
scheme issue statement	出图说明
scheme review/release meeting	审图/出图会议
science of combustion	燃烧学
science，technology，engineering and maths	科学、技术、工程、数学
scimitar-shaped leading edge	半月形前缘
scope clauses	条款范围
score	划痕

scramjet engine	超燃冲压喷气发动机
scramjet isolator	超燃冲压发动机隔离室
scratch	擦刮
screech	啸叫、振荡
screech/heat shield	隔热防护屏
screen printing	丝网印制
screening/editing/processing system for distortion data	畸变数据筛选/编辑/处理系统
screw pitch	螺距
screw propeller	螺旋推进器
scroll	蜗壳
Sea Harrier	海鹞-垂直起降战斗机(英宇航公司，英国)
sea level international standard atmosphere	海平面国际标准大气
seal clearance	封严间隙
seal pressure ratio	封严压比
seal pressurizing air	密封增压空气
sealant material	封严材料
sealing	密封
sealing element	密封元件
seals	密封件
seamless ring	无缝环形件
search and rescue	搜寻和营救
seatings for bolts	螺栓座
seats in standard layout	标准座位布局
second	秒
second derivative	二阶导数
secondary air	二次流
secondary air swirler	离心式空气喷嘴
secondary air system	辅助空气系统
secondary flows	二次流
secondary load path during fan blade off	风扇叶片分离时第二套传力线路
secondary oil/gas recovery	二次油气回收

secondary power system	用于军机地面服务的辅助动力系统（由辅助动力装置和两个齿轮箱组成）
secondary solidification structure	二次结晶组织
secondary structure	副承力结构
sector combustor test	燃烧室扇形段试验
securing rotor blades	固定转子叶片
see level	海平面
see level cabin	海平面座舱
see level take-off standard	海平面起飞标准条件
seed crystal	种晶
segment carrier	连接件
segmented CI023 nozzle	材料为 CI023 的扇形段喷嘴
self audit process	自审核流程
self directed team	自主团队
self-aligning coupling	自动定心联轴器
self-calibration	自校准功能
self-directed team	自主团队
self-independent innovation	自主创新
self-sealing housing	自封严壳体
self-sustain	自平衡
self-sustaining rpm	自持转速
self-sustaining speed	自平衡转速
self-testing	自测试
semiconductor discrete device	半导体分离器件
semiflexible wall	半柔壁
senate armed service appropriations committee	（美国）参院军事拨款委员会
sensitive altimeter	微动气压计
sensitive part	敏感件
sensor	传感器
sensor capacitance	传感器电容
separate jet	分离流排气
separate jet exhaust system	分离流喷气系统
separate jet plug core nozzle	分离流锥状中心喷管

separate nozzle	分流喷管
separate-flow nozzle	分离流尾喷管
SEPECAT Jaguar	美洲豹－战斗机(英宇航/布雷盖公司,英国/法国)
serial number	序列号
series production	批生产
series solution	级数解
Serious Fraud Office	(英国)严重诈骗事件调查办公室
serpentine cooled turbine	蛇管冷却涡轮
serrated nozzle	锯齿型喷管
service bulletin	服务通告(罗罗)
service ceiling	实用升限
service commitments	服务条款
service data manager	客户数据经理
service engineer	服务工程师
service engineer (S/E) 24/7 duty phone	每周7天每天24小时值班服务工程师电话
service hour	运营时数
service level agreement	服务水平协议
service life extension program	(美国空军)发动机使用寿命延长计划
service management plan	服务管理计划
service provider	服务商
service solutions for customer productivity	为客户的高效运营提供维护服务方案
serviceable condition	可使用状态
services packaging instruction sheet	包装服务规定表
servo control diaphragm	伺服控制膜片
servo controller	伺服控制器
servo metering valve	伺服计量阀
servo piston	伺服活塞
servo pressure	伺服压力
set screw	定位螺钉
set-up time	准备时间
several builds of core engine	几种核心机的组装机

severe bird strike requirement	严格的撞鸟要求
specific fuel consumption characteristic	发动机耗油率特性
shaft breakage	轴断裂
shaft buckling	纵弯失稳
shaft driven lift fan	轴驱动升力风扇
shaft horsepower	轴马力
shaft power	轴功率
shaft power engine	轴功率发动机
shaft power extraction	提取轴功率
shaft remote adjustment feature	涡轮间隙微调装置
shaft speed	轴转速
shaft strength analysis	轴强度分析
Shanghai Foreign Aviation Service Corporation	上海外航服务公司
Shanghai Technologies Aerospace Company Limited	上海科技宇航有限公司
shape memory alloy	变形记忆合金
shape-changing alloy	变形合金
share capital	股本
share in revenues	收益共享
shareholders' equity	股东权益
shear pin	剪力销
shear strength	剪切强度
sheet metal	钣金
sheet metal fabrication	钣金加工
sheet metal springback	金属板材成形回弹
shelf life code	储存期产品号
shelling	制壳
shelling process	脱壳流程
Shenyang Aeroengine Research Institute	沈阳航空发动机研究所
Shenyang Engine Design Institute	沈阳发动机设计研究所
Shenyang Liming Aero-Engines Group Corporation	沈阳黎明航空发动机集团公司
shielded metal arc welding	遮蔽金属电弧焊
shielded metal arc welding	手工电弧焊

shock absorber	减震器、阻尼器
shock cell jet noise	激波胞格射流噪声
shock dislocation	激波位错
shock formation	激波的形成
shock load	冲击载荷
shock noise	冲击噪声
shock wave	冲击波
shock-boundary interaction	激波与附面层相互作用
shop floor data management	车间数据管理
shop labour cost	维修工时费
shop manual	维修手册
shop material cost	维修材料费
shop visit	返修
shop visit rate	返修率
shop welding	车间焊接
Shore hardness	邵氏硬度
short circuit	短路
short circuit failure	短路失效
short duck separate flow	短涵道分离流
short duct nacelle	短涵道短舱
short live on wing	翼上寿命短
short take-off and landing	短距起降
short take-off and vertical landing	短距起飞垂直着陆
short take-off and vertical landing engine	短距起飞垂直着陆发动机
short take-off but arrested recovery	短距起飞,拦阻降落
short/medium range	中/短程
short-and medium-haul routes	中短航线
short-haul and medium-haul aircraft	中短程飞机
short-term win	短期收益
shot peening	喷丸
shoulder screw	带肩螺钉
shrouded 2 stage HP turbine design	二级带冠高压涡轮设计
shrouded dome	带遮蔽的喷嘴整流罩
shrouded propeller	罩筒螺旋桨

shrouded vanes	带外环的静子叶片
shroudless blade	无冠叶片
shrouds	封严环
shutdown	停车
shutdown alarm	停车报警信号
shut-off cock	停车开关
shut-off valve	截止阀,停车开关
Sichuan ChengFa Aero Science & Technology	(四川)成发科技公司
Sichuan Snecma Aeroengine Maintenance Corporation	四川斯奈克玛航空发动机维修有限公司
side link	吊挂侧向连接口
side maneuver	侧向机动
side-dump combustor	侧面突扩燃烧室
sideline	侧向噪声
side-mounted engines	侧装发动机
side-one maneuver	单侧动作
side-step maneuver	近距平行跑道着落操纵
sidewall compression inlet	侧压式进气道
Siemens Power Generation Group	西门子动力集团
Siemens-Westinghouse Power Corporation	西门子-西屋动力公司
significant in-service event	重大使用事件
significant mechanical damage	严重机械损伤
significant weight benefit	明显重量轻
silenced handling bleed valve	无噪声放气阀
silent aircraft	静音飞机
silicon carbide-reinforced titanium	碳化硅强化钛
Silvercrest	银冠发动机(赛峰集团,法国)
silver solder	银焊料
silvercrest	(赛峰计划中的)高性能支线飞机发动机
similarity conditions of compressor rig test	与压气机部件试验相似条件
simple bearing arrangement	简单轴承布局
simple cluster sampling	简单整群抽样
simple cycle	简单循环

simple random sampling	简单随机抽样
simulated altitude test facility	模拟高空试验设备
simulated altitude test of propulsion system	推进系统模拟高空试验
simulating plate	模拟板
simulation design technology	仿真设计技术
simulation deviation	模拟偏差
simulation model	仿真模型
simulator	模拟器
Singapore Aero Engine Services Limited	新加坡航空发动机服务有限公司
Singapore next fighter replacement program	新加坡战斗机更新项目
Singapore Technologies Aerospace Ltd.	新加坡宇航技术公司
singing propeller	蜂鸣推进器
single actuator	单一作动杆
single crystal	单晶
single crystal blade	单晶叶片
single crystal superalloy	单晶高温合金
single ended stator	单端面导向器
single multifuel nozzle	单极多种燃料喷嘴
single multi-valve block	多阀单体
single or double-sided impeller	单面或双面叶轮
single pass internal cooling	单通道内冷
single point detector	单点探测器
single sided entry combustor	单侧进气燃烧室
single spool turbojet	单轴涡轮喷气发动机
single status environment	不分等级的工作环境
single-annular tiled combustor	单级环形浮壁燃烧室
single-lever power control	单杆动力控制
single-phase valve	单向阀
single-piece bladed disk	整体叶片盘
single-rotation（propfan）	单转子(桨扇发动机)
single-shot oil pump	一次注油滑油泵
single-spool axial flow turbo-jet	单轴轴流式涡轮喷气发动机
single-spool compressor	单转子压气机

Sino-Swearing	华扬史威灵-商务飞机（华扬史威灵飞机公司，美国）
sintered	烧结成型
site rationalisation	地点合理化
situational assessment tool	（电话）监控软件
six-man test crew	6 人试飞组
size testing	尺寸测量
skew angle	叶片扭角
skills shortage	缺少技能和经验
Sky West	（美国）天西航空公司
skyraider	空中力量（无人机）
Skyteam	天合联盟
Skywatcher	空中哨兵（无人机）
slavaging and processing	修补和加工
slender fuselage	扁平机身
slide valve	分油活门
slinger	甩气盘
slipstream losses	滑流损失
slot and lug	键槽连接
slotted blade	带缝叶片
slotting	开槽
small airliner	小型客机
small bevel gear	小锥齿轮
small business innovation research	小型商务创新研究
small business jets	小型公务喷气机
small engine assembly line	小型发动机组装线
small engine component test facility	小发部件试验台
small freighter	小型货机
small heavy fuel engine	（霍尼韦尔公司）小型重油发动机
small unmanned air vehicle	小型无人机
small unmanned aviation vehicle engine	小型无人机发动机
small/medium enterprises	中/小型企业
smart component	智能部件
smart electronic scanning system	智能电子扫描系统

smart label	智能标签
smart materials	智能材料
smart sensors	智能传感器
smoke	烟雾
smoke number	烟雾指数
smooth airflow	平滑气流
snap-together final assembly	直接组装生产线
SNECMA	(法国)斯奈克玛公司
snout	进气嘴
snubber	叶间支撑
snubbered blade	带叶间支撑的叶片
Societe Nationale d'Etude et de Construction de Moteurs d'Aviation (Snecma)	斯奈克玛公司(赛峰集团,法国)
Societe Nationale Industrielle Aerospatiale (Aerospatiale)	(法国)法宇航公司
Societe Turbomeca	透博梅卡公司(赛峰集团,法国)
society of automotive engineers	汽车工程师协会
society of british aerospace companies	(英国)航空企业协会
society of japanese aerospace companies	(日本)航空企业协会
soft annealing	(使镍合金变软)软化退火
soft conversion	软性换算
soft fault of control system	控制系统软件故障
soft ignition	软点火(不产生瞬间压力突升)
softening temperature	软化温度
software configuration control board	软件配置管理委员会
software enabled control	软件启动控制
software implementing redundancy control	软件实现冗余度控制
software life cycle	软件寿命
solar power station	太阳能发电站
solar powerplant	太阳能动力装置
solar rocket engine	太阳能火箭发动机
solar thermal propulsion	太阳能火箭发动机
solar-powered aircraft	太阳能动力飞机
solder bump crack	焊点裂纹

solder joint fracture	焊点开裂
solenoid valve	电磁阀
solid backing sheet	实心底板
solid fuel ramjet vehicle	固体燃料超燃冲压飞行器
solid model	实体模型
solid oxide fuel cell	固态氧化物燃料电池
solid propellant	固体推进剂
solid rocket motor	固体火箭发动机
solid state bonding	固态连接
solid state coating technology	固态涂层技术
solid state power controller	固态功率控制器
solid titanium wide chord blades	实心宽弦钛合金叶片
solid wood packing material	实木包装材料
soluble core high pressure turbine	可分解的核心机高压涡轮
sonic boom	音爆
sonic boom effects	音爆效应
sonic fatigue	音波疲劳
sonic probe	声速探针
sonic velocity	声速
soot	积碳
soot formation	(燃烧室)积碳形成
sought-after buyer	很受欢迎的客户
sound absorption material	吸声材料
sound generation	声音的产生
sound intensity level	声强级
sound pressure contour	等声压线
sound pressure level	声压级
sound-proof	隔音
source	货源
source controlled drawing/definition	固定资源图/定义
source method control	源方法控制
sources for regulations	条例来源
sourcing	采购

Sovereig	奖状君主-公务机（塞斯纳飞机公司，美国）
space heat release rate	空间散热率
space-launcher	空间发射器
spacer	垫片
spanwise distribution of payload & fuel	翼展分布有效载荷和燃料
spare inventory	备用库存
spare parts	备件
spare parts availability	备件可获得性
spares account manager	备件客户经理
spark	电火花
spark ignition	火花塞点火
spatter	泼溅的污迹
SPE Aerosila JSC	（俄罗斯）空气动力联合股份公司
special electronic mission aircraft	特殊电子任务飞机
Special Federal Aviation Regulation	美国特种联邦航空条例
special process group	特定流程组
special purpose software for vibration analysis	振动分析专用软件
special purpose test equipment	专用试验设备
special security agreement	专门保密协议
special tools and test equipment	特殊工具和试验设备
special-purpose electrolytes glass beads	特种电解玻璃珠
specific	比
specific air range	比航程
specific enthalpy	比焓
specific flow function	流量函数
specific fuel consumption	耗油率
specific gravity	比重
specific heat	比热
specific heat ratio	绝热指数
specific load capacity	特定载荷量
specific power	特定功率
specific stagnation enthalpy	比滞止焓
specific thrust	单位推力

specific tool(s)	专用工具
specification search	（搜寻特定规格的产品）特寻
spectrochemical analysis	光谱化学分析
spectrometer	质谱仪
spectrometric oil analysis	滑油光谱分析
speed accuracy	速度精度
speed control	转速控制
speed control governor cam	转速控制调节凸轮
speed difference	转速差
speed fluctuation	转速波动
speed governor	传速调节器
speed hang-up	转速悬挂（即转速无法提高）
speed probe	转速探头
speed rollback	掉速
speed sensor	转速传感器
speed synchronizer	转速同步器
speedy business jet	快速公务机
Spey	斯贝发动机（罗罗公司,英国）
spill loss	溢损
spill valve	溢流阀
spillage drag	溢流阻力
spilling diffuser	溢流扩压器
spinner	顶锥
spiral bevel gear	螺旋锥齿轮
spiral groove gass face seal	螺旋槽端面气密封
Spirit	幽灵-轰炸机（诺斯洛普-格鲁曼公司,美国）
splash cooling strip	折流冷却板
spline coupling	花键联轴节
spline design	花键设计
spline grinding machine	花键磨床
spline hub	花键座
spline key	花键
spline shaft	花键轴

splines	样条曲线
split manifold	分束输油管
split ring	扣环
splitter	分流器
splitter fairing	分流器整流罩
spontaneous ignition	自燃
spool speed	转子速度
spool valve	柱形阀
spray bar	喷油管
spray field	喷雾场
spray metal molding	金属喷涂模
spray nozzle	燃油喷嘴
spread	分布
spring bearing	弹簧轴承
spring clip	弹簧夹
spring lock washer	弹簧锁紧垫圈
spur gear	直齿轮轴
spur gear shaft	直齿轮轴
spurious grains	伪晶
spy mission	间谍任务
square external	外方口螺母装卸套
square female	方口螺母
square internal input drive	方形内口驱动套
square socket	方口套筒
square to face	垂直于表面
squeeze film	挤压油膜
squeeze film and integral squirrel cage	挤压油膜与一体化鼠笼支撑
squeeze film bearing	挤压油膜轴承
squeeze film damper	挤压油膜阻尼器
squeeze film outer race	挤压油膜外环
squirrel cage motor	鼠笼式电机
squirrel cage support	鼠笼式弹性支撑
ST Aerospace	新加坡科技宇航公司
stability	稳定性

stability analysis technique	稳定性分析技术
stability assessment factor	稳定性评定系数
stability assessment verification test	稳定性评定验证试验
stability line	稳定线
stabilizer	稳定器
stage 2 stator	第二级静叶
stage 4 noise regulations	第四级噪声标准
stage inspection	阶段检验
stage length	航段长度
stage pressure ratio	级增压比
staged combustion technology	分级燃烧技术
staged combustor	分级燃烧室
stage-V	第五阶段标准
stagger angle	安装角
stagnant air	气流停滞
stagnation enthalpy	制止焓
stagnation pressure	驻点压力
stagnation temperature	驻点温度
stainless steel	不锈钢
stall	失速
stall free	无喘振
stall free operation	（发动机）无喘振运行
stall limit	失速极限
stall margin	失速裕度
stall speed	失速
stall torque	失速扭矩
stalling speed	失速速度
standard annular combustor	标准环型燃烧室
standard atmosphere	标准大气
standard cubic meter	标准立方米
standard data records office	标准数据记录办公室
standard deviation	均方差
standard generalised markup language	标准通用标示语言
standard heat value	标准热值

standard inspection procedure	检验标准程序
standard instrument departure	标准仪表离场
standard operations management system	标准运营管理系统
standard tools and equipment	标准工具与设备
standby engine	备用发动机
standby fuel pump	备用燃料泵
standby generator	备用发电机
standby power system	备用电源系统
standing operating procedure	标准操纵规定
Star Alliance	星空联盟
star flex	爪型联轴器
starboard	右舷
starboard engine	右侧发动机
starboard wing	右翼
start and landing permission	起动和着陆许可
start control system	起动控制系统
start process	起动过程
start schedule	起动计划
start time	起动时间
starter	起动机
starter air control valve	启动器气控阀
starter air valve	起动机气阀
starter assistance	起动机辅助
starter jaw	起动机棘爪
starter mount	启动器安装架
starter valve open	启动器阀门打开
starting	起动
starting drive train	起动传动齿轮系
starting ignition device	起动点火器
starting limit	起动极限
starting system	起动系统
start-up course	启动过程
start-up pressure difference	起动压差
start-up pump	启动泵

state queue of aero engine fleet	发动机机队状态排序
state subsidies	国家补贴
statement of requirements	需求报告
statement of work	任务说明书
state-of-the-art technology	最新技术
state-run carriers	国有航空公司
state-run company	国有公司
static analysis	静力分析
static balance	静平衡
static casing stiffness test	静力机匣刚度试验
static pressure	静压
static reduction	静力缩减（自由度）
static seal segment	非转动环形件
static strength of structures	结构静强度
static tapping	静压侧孔
static temperature	静温
stationary gasturbine gen set	固定式燃机发电机组
stationary plasma thruster	静态等离子推进器
stations	截面
statistical process control	统计过程控制
stator outlet temperature	静子出口温度
stator outlet temperature	涡轮前导向器出口温度
stator outlet temperature comparison	涡轮前导向器出口温度比较
stator shroud leakage	静子叶冠渗漏
stator throat width	静叶喉部宽度
stator vane actuator	导向器作动器
stator without bow	无扭曲静子叶片
statorless tubine	无导向器涡轮
stator-rotor shock interaction	静子和转子激波的相互作用
status monitoring and failure diagnosis	状态监控与故障诊断
steady state engine-speed measurement	稳态发动机转速测量
steady state forced response	稳态强迫响应
steady state performance study	稳态性能分析
steady state performance test	稳态性能试验

stealth	隐形
stealth inlet	隐身进气道
stealthy technology	隐身技术
steam generator	蒸汽产生器
steam heater	蒸气加热器
steam turbine generator	蒸气涡轮发电机
steel (SCMV) shaft	钢轴(钢牌号为 SCMV)
steel casing	钢机匣
steel die forgings	钢模锻件
steel metal shroud segment	钢制叶冠
steep approach landing	大俯角进近着陆
steep inclined combustor	通道陡斜的燃烧室
steep low pressure turbine annulus	陡斜低压涡轮流道
stellite	司太立硬质合金
stellite overlay melding	司太立硬质合金堆焊
stem drilling	深孔钻削
step climb rating	阶梯爬升状态
step-aside gearbox	测装齿轮箱
step-change technologies	变革性技术
stereotithography	光敏树脂成形
stereotithography apparatus	光敏树脂快速成型机
stiff high pressure compressor casing structure	高刚度高压压气机机匣结构
stiff rotor	刚性转子
stirrer	搅拌器
stock certificate	股票
stockholders' equity	股东权益
stoichiometric ratio	化学计量比
stopway	停止道
storage bag	(发动机)储存袋
stow or deploy movement	开启或关闭过程
straight hole	直孔
straight tooth spur gear	直齿正齿轮
straight-line-moving hydraulic actuator	直线运动液压作动器
straight-through combustor	直通式燃烧室

straight-through flow system	直流气流系统
strain gauging	应变片测量
strainer	滤网,过滤器
straingauge	应变片
stranger & alien style components	异型风格部件
strategic alignment	战略联合
strategic alliance	战略联盟
strategic research agenda	(欧盟)战略研究计划
strategic sourcing group	战略采购组
strategic sourcing initiative	战略采购方案
strategic sourcing supplier selection	战略采购供应商的选择
strategy quality assurance	策略质量保证
stratolaunch	(美国)同温层发射(巨型飞机)
stratosphere	同温层
stratospheric surveilance airship	同温层侦察飞艇
stream traces	气流轨迹
streamtube engine	流管发动机
strength and life statement	强度和寿命的说明
strength life and integrity statement	强度寿命及完整性说明
strength of tip shroud	叶冠强度
strengths, weaknesses, opportunities, threats	强项,弱项,机会,威胁
stress corrosion fracture	应力腐蚀断裂
stress in disc	盘应力
stress rupture life	持久寿命
stress rupture property	持久性能
stress/vibration analysis	应力/振动分析
Strike Master	打击能手-教练机(英宇航公司,英国)
strike by lightning	雷击
stripped down and rebuilt	(发动机)分解和重组
stroke	冲程
stroke throttle valve	行程节流阀
structural by-pass duct	承载结构外涵道
structural damping	结构阻尼
structural distortion	结构变形

structural health monitoring	结构健康检测
structural steel	结构钢
structural strength	结构强度
structural test	结构试验
structural validation	结构验证
structural vibration loads	结构振动载荷
structure fatigue test	结构疲劳试验
structural-thermal coupling	结构热耦合
structured reactive	结构化反应
structural integrity	结构完整性
strut	承立杆
stub shaft location bearing	短轴定位轴承
stud bolt	无头螺栓
S-type inlet	S型进气道
sub-contract manufacture	转包生产
sub-contractor	分包商
submarine rescue system	水下救生系统
submarine rescue vehicle	水下救生艇
sub-scale model turbine test	涡轮模型试验
subsonic fixed wing	亚声速固定翼
subsonic flow	亚声速流
subsonic gas turbine engine	亚声速涡轮发动机
subsonic inlet	亚声速进气道
subsonic stage	亚声级
substances of very high concern	(欧盟)高度担忧物质
substantiation plan	验证计划
subsurface imaging radar	地面成像雷达
sub-system accountability	子系统问责制
sub-system definition document	子系统定义文件
sub-system requirements document	子系统需求文件
subsystem supplier	子系统供应商
sub-tier performance	基层业绩
sub-tier supplier	次级供应商
sub-tier supplier management	次级供应商管理

succeeding stages	后续级
success criteria	成功的要求
successful handling point	成功的放气点
suction face	负压面
suction pressure	吸附压力
suction side	(叶片)吸力面
suction strainer	吸入过滤器
suction valve pressure	吸流阀压力
suitably qualified and experienced personnel	有资质和经验的员工
Sukhoi Aviation Holding Company	(俄罗斯)苏霍伊航空集团
Sukhoi Civil Aircraft Corporation	(俄罗斯)苏霍依民机公司
Sukhoi Design Bureau	(俄罗斯)苏霍伊飞机设计局
sulphidation resistent	抗硫性
sulphur dioxide emission	二氧化硫排放
sump oil debris sensor	油箱碎屑检测器
super alloy	高级合金
Super Etendard	超级军旗-攻击机(达索公司,法国)
Super Fan	超扇发动机
Super Jet	超级喷气-支线机(苏霍伊设计局,俄罗斯)
Super Lynx	超山猫-直升机(韦斯特兰与奥古斯特合资,英国/意大利)
super maneuverability	超机动性
Super Mystere	超神秘-战斗机(达索公司,法国)
super plastic forming	超塑成型
super plastic forming titanium fairing	超塑成型钛合金整流片
Super Skyhawk	超级天鹰-攻击机(洛克希德-马丁公司,美国)
superalloy ingot	高温合金锭
supercharger	增压器
supercooled water droplets	过冷水珠
supercritical	超临界
supercritical airfoil	超临界叶型
supercritical running range	超临界运转范围

supercritical wing	超临界机翼
super-efficient	超高效
superfreighter	超大型货机
Superjet	(俄罗斯)超级喷气支线客机
super-plastically formed	超塑成形
superposition method	叠加算法
supersonic	超声速
supersonic airflow	超声速气流
supersonic and transonic nozzle	超跨声速喷管
supersonic business jet	超声速公务机
supersonic combustion	超声速燃烧
supersonic combustion ramjet	超燃冲压喷气发动机
supersonic combustion scramjet engine	超燃冲压发动机
supersonic concorde	超声速协和式客机
supersonic cruise	超声速巡航
supersonic cruise aircraft research	超声速巡航飞机研究
supersonic gas turbine engine	超声速涡轮发动机
supersonic inlet	超声速进气道
supersonic inlet air velocity	超声速进口空气速度
supersonic stage	超声级
supersonic through-flow stage	超声速通流级
supersonic transport	超声速运输机
supersonic turbine	超声速涡轮
supertanker	超级油轮
supplementary firing	补充燃料
supplementary intake	辅助进气口
supplier advanced business relationship	(罗罗公司与)供应商的先进业务关系
supplier briefing pack	供应商须知
supplier delivery specification	供应商交货规范
supplier development & quality	供应商的开发与质量
supplier development improvement leader	供应商发展改进负责人
supplier development leader	供应商发展负责人
supplier furnished equipment	供应商安装的设备
supplier information system	供应商信息系统

supplier management team	供应商管理小组
supplier progress report	供应商进展报告
supplier quality memorandum	供应商质量备忘录
supplier quality renaissance	供应商质量复兴项目（罗罗）
supplier total evaluation process	供应商全面评估流程
suppliers chain management	供应链管理
supply chain capability	供应链能力
supply chain event management	供应链业务管理
supply chain integration team	供应链一体化小组
supply chain logistics	供应链物流
supply chain operations centre	供应链运营中心
supply chain planning capability	供应链计划能力
supply chain relationships in aerospace	航空航天工业中的供应链关系
supply chain strategy	供应链战略
supply chain unit	供应链管理部
supply quality engineer	供货质量工程师
supply quality team	供货质量组
support stiffness	支承刚度
supportive Infrastructure	支援性基础设施
suppressing noise	抑制噪声
suppressor nozzle	消音喷管
surface area	表面积
surface air cooled oil cooler	表面气冷滑油冷却器
surface cloudburst	表面喷丸处理
surface combatant maritime rotorcraft	（英国）水面作战海上旋翼机
surface element and valence analysis	表面元素及化合价分析
surface finish	表面处理
surface hardening	表化硬化
surface micro-area analysis	表面微区分析
surface pressure distribution	表面压力分布
surface profilometer	表面光度仪
surface resistivity	表面电阻率
surface-to-air missile	地空导弹
surge	喘振

surge limit	喘振边界
surge line	喘振线
surge line threat	喘振边界的风险
surge map	喘振特性图
surge margin	喘振裕度
surge margin deterioration	喘振裕度衰减
surge margin retention	喘振裕度保持
surge margin stack-up	喘振裕度累积
surge pressure ratio loss	喘振压比损失
surplus fuel	残余的燃油
surveillance of production	生产监控
surveillance report	监视报告
surveillance review	监督评价
suspension bearing rotor	悬浮轴承转子
sustainable cost	可承受的价格
sustainable aviation fuel	(绿色环保)可持续航空燃料
sustained fire	持续的火情
sustaining engineering	持续工程
swan necked duct	鹅颈状流道
swarf	碎屑
Swedish Defence Material Administration	瑞典国防装备管理局
Sweeney	(美国发动机维修工具制造商)斯威尼公司
Sweeney tools	斯威尼工具
sweep blade	后掠叶片
swept fan	后掠风扇
swept fan outlet guide vanes	后掠风扇导向叶片
swept rotors	后掠转子
swept-back delta wing	后掠三角翼
swinging angle	摆角
swing-wing	可变后掠翼
swirl afterburner	涡流加力燃烧室
swirl angle	涡流角
swirl atomizer	涡流式喷油嘴

swirl augmentor	涡流加力燃烧室
swirl cup design	涡流式杯型设计
swirl generator	涡流发生器
swirl nozzle	涡流喷嘴
swirl number	涡流数
swirl type combustor	涡流形燃烧室
swirl vanes	涡流导向叶片
swirl-can combustor	涡流罐式燃烧室
swirler	涡流器
swirl-venturi fuel injector	涡流—文氏管形喷油器
switched reluctance motor	开关磁阻电机
switch-in deflector	开关式偏转器
switch-in deflector system	开关式偏转器系统
swivelling nozzle	旋转喷管
synchronizing	同步
synchronous generator	同步发电机
synchronous idle	同步慢车
synchronous speed	同步速度
synchrophasing	同相
synthesis	合成
synthesis distortion coefficient	综合畸变指数
synthesis gas	人造然气(德国人在 1923 发明)
synthetic aperture radar	合成孔径雷达
synthetic oil	合成滑油
syntroleum	(从天然气转化的)合成燃油
system and mechanical tests	系统和机械试验
system anomaly note	系统异常记录
system application product	系统应用产品
system configuration management plan	系统方案管理计划
system design and demonstration	系统设计和验证
system design and development	系统设计和开发
system design and unit requirement definition	系统设计和零件要求定义
system design process	系统设计流程
system development and demonstration	(F135 发动机)系统发展和验证

system integration	系统集成
system integrator	系统集成商
system packaging	系统集成
system program office	系统计划办公室
system reconfiguration	系统重新构型
system verification vehicle	系统验证机

tacholimiter	转速限制器
tachometer	转速表
tachosensor	转速传感器
tack welding	点焊
tactical airlifter	战术运输机
tactical fighter experimental	战术战斗机实验机
tactical transport aircraft	战术运输机
tactical transport helicopter	战术运输直升机
tactical unmanned air vehicle	战术无人机
tag information management system	标识信息管理系统
Taikoo（Shandong）Aircraft Engineering Co.，Ltd.	山东太古飞机工程有限公司
tail bearing housing	尾部轴承架
tail reduction gearbox	尾部减速器
tail rotor driving shaft	尾桨传动轴
tail rotor pitch control	尾桨变距控制
tailored arrival	定制进场
tailored fibre placement	定制的纤维铺放
tailor-made blading	定制叶型
take off weight	起飞重量
take-off/take off	起飞
takeoff clearance	起飞许可
take-off decision speed	起飞决断速度
take-off distance	起飞距离

take-off field length	起飞跑道长
takeoff overall pressure ratio	起飞时整机压比
take-off rating	起飞推力设置
takeoff roll	起飞滑跑
take-off safety speed	安全起飞速度
take-off/landing slot	起降空档
Talos	(美国)塔洛斯导弹
tandem blade	串列叶片
tandem cascade	串列叶栅
tandem fan vectored thrust engine	串列风扇推力矢量发动机
tangential blowing blade	切向喷气叶片
tangible goods	有形货物
TAP Air Portugal	葡萄牙航空公司
tap bolt	自攻螺栓
tape-laying machine	铺带机
tapping screw	自攻螺钉
Taranis	雷神-无人机(英宇航公司,英国)
target detection system	目标探测装置及电子系统
target drone	靶机
target gap	目标差距
tariff barrier	关税壁垒
tariff concession rate	关税优惠率
tarmac	停机坪
task force	特别工作组
task control architecture	任务监控方案
TATA Advanced Material Limited	(印度)塔塔先进材料有限公司
tax credit	减税
tax on transaction	交易税
tax on value added	增值税
tax refunded	已退回的税款
tax-free trade zone	免税贸易区
taxi-in time	滑行入场时间
taxi-out time	滑行出场时间
Tay	泰发动机(罗罗公司,英国)

team based working	团队化工作方式
Tech 56	(CFM 公司) CFM56 发动机新技术开发计划
Techjet Aerofoils Ltd (TechJet)	(以色列)泰捷航空叶片公司
technical capability review	技术能力评审
technical corporate support	母公司技术支持
technical data integration	技术参数一体化
technical data tracking monitor	技术数据跟踪监控
technical finding	技术问题
technical information management system	技术信息管理系统
technical instruction	技术手册
technical life review	工程寿命评审
technical note	技术说明
technical oversight assessment	技术监督评估
technical oversight assessment-planning & control	计划与控制的技术监督评估
technical oversight assessment-quality	质量的技术监督评估
technical publication	技术出版物
technical readness level	技术成熟水平
technical report	技术报告
technical review	技术评审会
technical service directory	技术服务手册
technical specification	技术规范
technical support & operations	技术支持和经营
technical surveillance metallurgist	冶金技术监控专家
technical variance	技术超差
technical variance	技术变更
Technocomplex	(俄罗斯)技术生产中心(公司)
technological breakthrough	技术突破
technological heritage	技术的继承性
technology breakthrough	技术突破
technology concept airplane	技术方案验证机
technology endurance test	工艺耐久性试车
Technology for Advanced Low NOx	(普惠公司的)"泰龙"燃烧室

technology readiness program	技术准备项目
technology trend	技术发展趋势
Techspace Aero	技术空间航空技术公司(赛峰集团,比利时)
Tedlar	(杜邦公司生产的)聚氟乙烯薄膜
Tefzel	(杜邦公司生产的)新型氟聚合物
Teledyne-Ryan	特里达因-瑞安公司(诺斯罗普-格鲁曼,美国)
telemetering and telecontrol	遥测遥控
telemetry	遥感测量
telemetry equipment and related technologies	遥测设备及其技术
temperature	温度
temperature and pressure variation through turbine	涡轮温度和压力变化
temperature indicating paints	示温涂料
temperature limiter	温度限制器
temperature phial	温度计
temperature profile	温度分布
temperature rise ratio	加温比
temperature sensor	温度传感器
temporary guidance leaflet	临时指导单
tenon strength	榫头强度
tensile modulus of elasticity	拉伸弹性模量
tensile strain at break	断裂拉伸应变
tensile strength	拉伸强度
tensile stress-strain curves	拉伸应力-应变曲线
tentative agreement	意向性协议
terminal post	接线柱
terminal shock	边界激波
terms of business	商务条件
terrain clearance	离地高度
test bed	试车台
test bed analysis	试车台分析
test bed cross-calibration	试车台交叉校准

test bed thrust measurement	台架推力测量
test data analysis	试验数据分析
test data playback program	试验数据回放程序
test facility schematic	试验设施简图
test process instruction	测试工艺
test request	试验要求
test requirements document	试验要求文件
testing failure	试验失效
testing high-tech objectives in reality	现实中的高科技目标实验(空客 3D 打印飞机)
testing on air and oil system	油气系统的试验
Texas Aero Engine Services Limited	得克萨斯航空发动机服务有限公司
text element identifiers	文本单元标识符
Textron Lycoming	(美国)德事隆·莱康明公司
texture	材料结构
the City	伦敦股票交易所
the City code	伦敦股票交易所代码
theoretical size	理论正确尺寸
theory of inventive problem solution	发明解决问题的理论
thermal anti-icing	热除冰
thermal barrier coating	热障涂层
thermal battery	热电池
thermal bimetal	热双金属
thermal conductivity	热传导
thermal efficiency	热效率
thermal electrical equipment	热电设备
thermal expansion	热膨胀
thermal expansion of turbine casing	涡轮机匣热胀
thermal fatigue	热疲劳
thermal gravity analysis	热重分析法
thermal insulation	隔热
thermal load	温度载荷
thermal matching of rotor and casing	转子和机匣热匹配
thermal paint	示温漆

thermal resistance thermometer	热电阻温度计
thermal shock resistance	抗热震
thermal shock test of blade	叶片热冲击试验
thermal spray coating	热喷涂层
thermal stress	热应力
thermal wave imaging	热波成像
thermally matched	热匹配
thermal-mechanical duty	热力载荷
thermal-mechanical fatigue	热应力疲劳
thermochemical process	热化学工艺
thermocouple	热电偶
thermocouple pyrometer	热电偶高温计
thermocouple thermometer	热电偶温度计
thermodynamic cycle	热力循环
thermodynamic processes	热力过程
thermodynamic state	热力学状态
thermodynamic system	热力学体系
thermometer	温度计
thick thermal barrier coating	热障厚涂层
thin section	薄壁部分
thin solid blading	薄实心叶片
third party liability	第三者责任
thread form	螺纹形式
thread-type oil filter	螺纹式滑油滤
three-axel sensor	三向传感器
three-dimensional flow	三维流体
three-dimensional inlet	三维进气道
three-shaft dynasty	三轴发动机王朝
throat	进气道喉部
throttle by-pass adjuster	油门旁路调节器
throttle characteristics	油门杆特性
throttle element	节流元件
throttle grips	油门杆手柄
throttle lever	油门杆

throttle lever position in transition state	过渡状态油门杆位置
throttle movement	油门杆的移动量
throttle outlet pressure	油门出口压力
throttle resolver angle	推力杆解算器角度
throttle transient	推力杆瞬时状态
throttle unit	油门装置
throttle valve	油门开关
throttle ventilation	节流通风
throttles only control	仅靠推力杆控制
through life support	全寿命期支持
thrust	推力
thrust balance	推力平衡
thrust class	推力等级
thrust control malfunction	推力控制故障
thrust derate	减推力
thrust lever angle	推力杆角度
thrust link	推力传递杆
thrust load	推力载荷
thrust measuring rig	推力测试台
thrust per frontal area	单位迎面推力
thrust performance	推力性能
thrust rating	额定推力
thrust response	推力响应
thrust restoration	推力恢复
thrust reverser	反推力装置
thrust reverser operation	反推装置运转
thrust reverser actuation system	反推力作动系统
thrust reverser control unit	反推力控制器
thrust reverser shown deployed	反推装置打开的状态
thrust reverser unit	反推力装置
thrust ring	承推环
thrust strut	推力承载杆
thrust strut attachment lug	承推杆安装环
thrust to weight ratio	推重比

thrust、efficiency、new technology	推力、效率、新技术（罗罗）
thrusters	推进器
thrustmeter	推力计
Thunderbolt	雷电-攻击机（仙童公司,美国）
Ti alloy bar	钛合成棒
Ti extrusion	挤压钛件
tidal generator	潮汐发电机
tidal power	潮汐发电
tie rod	联杆
tied-shaft rotor	联轴转子
Tiger	虎式-战斗机（诺斯洛普-格鲁曼公司,美国）
time between overhauls	翻修寿命
time domain analysis of dynamic data	动态数据时域分析
time expiry	期满
time limited dispatch	限时遣派
time limits manual	时效手册
time-control capability	进度控制能力
timed fuel pump	定时燃油泵
timing hole	调整孔
tip clearance	叶尖间隙
tip clearance control	叶尖间隙控制
tip clearance distortion	叶尖间隙干扰
tip clearance variation	叶尖间隙变化
tip leak	叶尖泄漏
tip region	叶尖区
tip speed	叶尖速度
tip vortex	叶尖涡
tip-turbine	叶尖涡轮
tip-turbine driven lift fan	叶尖涡轮驱动升力风扇
tip-turbine fan concept	叶尖涡轮风扇方案
tip-vortex-free propeller	无尖涡螺旋桨
titanium alloy	钛合金
titanium alloy metal matrix composite	钛合金金属基复合材料

titanium aluminide	铝化钛
titanium billet	钛锻坯
titanium blisk	钛叶片盘
titanium blisk design	钛叶片盘设计
titanium casing	钛机匣
titanium castings	钛铸件
titanium conventional blade and disk	常规钛叶片和盘
titanium fairing	钛合金整流罩
titanium fan disc	钛合金风扇盘
titanium fire prevention coating	钛防火涂层
titanium fire protection layer	钛防火层
titanium forgings	钛铸件
titanium ingot and billet	钛锭和钛坯
titanium matrix composite	钛合金金属基复合材料
titanium seamless rings	无缝钛环件
titanium silicon carbide composite bladed ring	含钛碳化硅复合材料叶片环
titanium welded drum with titanium blading	带钛合金叶片的焊接钛盘鼓
titanium-graphite	石墨纤维钛合金层板
titanium-stabilized duplex stainless steel	钛稳定双相不锈钢
titanium cadmium coatings	钛镉涂层
to be advised	待通知
to be determined	待定
tolerance of fasteners	紧固件工差
Tomcat	雄猫-战斗机(诺斯洛普-格鲁曼公司,美国)
tonnage plan	吨位订单计划
top dead centre	顶部驻点
top off climb	最高爬升点
top tier airlines	最佳航空公司
topology	拓扑学
topology optimisation	拓扑优化
TopSolid	一种计算机辅助设计和制造一体化软件

Toray	东丽公司生产高强碳纤维环氯树脂材料
torgue density	扭矩载荷密度
torn	裂缝
Tornado	狂风-战斗机（德宇航/英宇航公司，德国/英国）
Tornado combat aircraft	狂风战斗机
toroidal vortex	环状涡流
torque delta	扭矩增量
torque density	扭矩密度
torque motor	扭矩马达
torque multiplier	扭矩放大器
torque pressure transmitter	扭转压力转换器
torque ratio	扭矩比
torque tightening	扭矩拧紧
torque tooling	扭矩工具
torque wrench	扭矩扳手
torque-measuring mechanism	测扭矩机构
torquemeter pump	扭矩计油泵
torsion meter	扭矩表
total air temperature	空气总温
total assets	总资产
total balance	总计余额
total budget	总预算
total care	（罗罗公司发动机）全面维护
total care agreement	（罗罗公司发动机）全面维护协议
total care package	包修协议包
total commitment	承担总值
total component support	（汉莎技术公司的）全面部件支持
total engine support	（汉莎技术公司的）全面发动机支持
total equity	权益合计
total fixed nitrogen	（燃气分拆）全部固态氮
total global spend	全球总花费
total inlet air pressure	进气道空气总压

total inlet air temperature	进气道空气总温
total inventory	总库存量
total liabilities	负债合计
total office management	办公室全面管理
total operating expenses	营业费用总计
total operational support	全面运营支持
total pressure recovery coefficient	总压恢复系数
total productive maintenance	全面生产性维护
total quality control	全面质量控制
total quality management	全面品质管理
total stockholders' equity	股东权益总计
total tangible assets	有形资产总计
total technical support	(汉莎技术公司的)全面技术支持
total threat	总体威胁
total weight of fuel	总燃油量
total workforce	雇员总数
TotalCare	(罗罗公司发动机)全面维护协议
touch down	着陆
toughened epoxy matrix	强化环氯基
tower shaft	锥齿轮轴
tower shaft input	锥齿轮轴输入
toxic air	有毒气体
track lock valve	轨道保险阀
trade-off analysis	权衡分析
trade-off study	利弊分析
traffic collision avoidance system	空中交通防撞系统
trailing edge	(叶片)后缘
training and logistics support	培训和后勤支援
trajectory test	轨迹试验
trans-Atlantic northern-Pacific	跨大西洋北太平洋
transfer efficiency	低压效率
transient	瞬态
transient distortion	瞬时畸变
transient engine operation	瞬时发动机工作状态

transient liquid phase	瞬间过渡液相
transient manoeuvres	瞬时机动动作
transient overshoot	瞬时复飞
transient performance simulation	瞬时性能模拟
transient performance study	瞬态性能分析
transient speed measurement	瞬态转速测量
transient temperature distortion	瞬时温度畸变
translating cowl	平移罩
transmission efficiency	传动效率
transmission system	传动系统
transonic compressor	跨声速压气机
transonic fan rotor	跨音速风扇转子
transonic flow	跨音速流
transonic stage	跨声级
transonic turbine	跨声速涡轮
trans-Pacific	跨太平洋
transpiration cooling	蒸发式冷却
transpiration cooling blade	蒸发式冷却叶片
transply combustor	多孔层板发散冷却燃烧室（罗罗）
Transport Canada	加拿大运输部
Transport Canada Civil Aviation	加拿大运输部民航局
Transportation Security Agency	（美国）运输安全局
transportation stand	（发动机）运输托架
transportation worker identification credential	（美国）运输工人身份卡
transverse vibration	横向振动
trapped volume	（齿轮泵）齿间集油容量
trapped vortex combustor	驻涡燃烧室
traverse temperature control	截面温度控制
trembler mechanism	断续器机构
Trent	遄达发动机（罗罗公司,英国）
trial and error method	试错法
trichloroethane	三氯乙烷
tricycle	三点式
Trident	三叉戟-客机（英宇航公司,英国）

trident missile system	（英国）三叉戟潜艇的导弹系统
triggered assessment	触发性评估
triggered isomer heating exchanger	（美国）触发同分异构体热交换器
trim balance probe	平衡探头
trimmer actuator	微调作动器
trims	微调
triple spool compressor	三转子压气机
triple-breech cartridge starter	三筒式火药筒起动机
TriStar	三星-客机（洛克希德-马丁公司,美国）
Triton	海神-无人机（诺斯洛普-格鲁门公司,美国）
Triumph	凯旋-公务机（比奇飞机公司,美国）
tropopause	对流顶层
troposphere	对流层
troposphere wind profiling	对流层风廓线
trouble free engine	无故障发动机
trouble shooting	排故
trouble-free operation	无故障运营
true air speed	真空速
true ground speed	真地速
truncated	截短的
TU Darmstadt Department of Gasturbines and flight Propulsion	达姆施塔特工业大学院燃气涡轮和飞行推进研究所
TU Dresden Institute of Fluidmechanics	德累斯顿工业大学流体力学研究所
TU München Institute of Flight Propulsion	慕尼黑工业大学飞行推进研究所
tubeless vortex reducer	无管道涡旋导流器
tubo-annular combustor	环管燃烧室
tubular combustor	管道燃烧室
tubular rivets	空心铆钉
tuition session	辅导课
tumble vortex	滚涡
tun-dish	浇口盘
tungsten alloy	钨合金

tungsten inert gas welding	惰性钨气保护焊
tuning of mount stiffness	调节吊架刚度
Tupelev Experimental Design Bureau	(俄罗斯)图波列夫实验设计局
Tupelev PJSC	(俄罗斯)图波列夫联合股份公司
Tupolev Design Bureau	(俄罗斯)图波烈夫设计局
turbine	涡轮
turbine active clearance control	涡轮间隙主动控制
turbine adiabatic efficiency	涡轮绝热效率
turbine aerofoils	涡轮叶片
turbine based combined cycle	涡轮基联合循环(发动机技术)
turbine blade cooling effectiveness test	涡轮叶片冷却效果试验
turbine blade seal segment	涡轮叶片封严圈
turbine blade thermal fatigue test	涡轮叶片热疲劳试验
turbine blade triangle	涡轮叶片速度矢量三角形
turbine bypass engine	涡轮涵道发动机
turbine cascade	涡轮叶栅
turbine casing	涡轮机匣
turbine casing mounting flange	涡轮机匣安装边
turbine cooling air	涡轮冷却空气
turbine cooling system	涡轮冷却系统
turbine discharge pressure	涡轮出口压力
turbine efficiency	涡轮效率
turbine endwall profile	涡轮壁端截面
turbine entrance	涡轮入口
turbine entry temperature	涡轮进口温度
turbine exhaust case	涡轮排气机匣
turbine exit	涡轮出口
turbine exit case	涡轮出口机匣
turbine exit vane	涡轮出口整流器
turbine fairing	涡轮整流罩
turbine flow meter	涡轮流量计
turbine flow separation	涡轮气流分离
turbine gas temperature	涡轮燃气温度
turbine genset package	涡轮发电机组

turbine inlet temperature	涡轮进口温度
turbine intermediate case	涡轮中介机匣
turbine map	涡轮特性图
turbine nozzle	涡轮导向器
turbine outlet passage pressure loss	涡轮出口流道压力损失
turbine overspeed	涡轮超速
turbine overtip leakage	涡轮叶尖漏损
turbine passages	涡轮流道
turbine performance	涡轮性能
turbine rotor	涡轮转子
turbine rotor blade	涡轮转子叶片
turbine temperature probe	涡轮温度探头
turbine work per mass of air flow	单位流量涡轮功
turbine work split	涡轮功分配
turbo electric propulsion	涡轮机电力推进
turbo fuel pump	涡轮燃油泵
turbo water pump	涡轮水泵
turbocharge engine	涡轮增压发动机
turbocharge impeller	涡轮增压器叶轮
turbo-driven compressor	涡轮压缩机
turbo-driven supercharger	涡轮增压器
turbofan	涡扇发动机
turbofan engine	涡扇发动机
turbojet engine	涡喷发动机
turbomachinery flow fields	叶轮机流场
turbomachinery gas dynamics	涡轮机气动力学
Turbomeca	(法国)透博梅卡发动机公司
Turbomeca Africa Pty	透博梅卡直升机发动机非洲公司
turbo-prop	涡轮-螺旋桨
turboprop engine	涡轮螺旋桨发动机
turbo-prop engine	涡桨发动机
turbopump	涡轮泵
turbo-rocket engine	涡轮-火箭发动机
turboshaft engine	涡轴发动机

Turbo-Union Ltd.	涡轮联合有限责任公司
Turbo-Union，Bristol	涡轮联合公司布里斯托尔分部
turbulated holes	扰流孔
turbulence	紊流,湍流
turbulence generator	湍流发生器
turbulence intensity	紊流度
turbulence ring	紊流环
turbulent flow	紊流
turbulent separation	紊流分离
turn around time	周转时间
turn round time	调头时间
turnaround time	完成时间
turning and milling operations	车和磨工艺
turnkey project	总承包项目,交钥匙项目
twin nozzle deflector	双喷管偏转器
twin spool gasturbine	双轴燃气涡轮机
twin spool mixed turbofan	双轴混合流涡扇发动机
twin thrust strut	双推力杆
twin-aisle airliner	双通道客机
twin-aisled dreamliner	双通道梦幻客机(波音 787)
twin-annular premixing-swirler	双环腔油气预混涡旋(GE)
twin-annular premixing-swirler combuster	双环腔预混涡流燃烧室(GE)
twin-deck airliner	双层客机
twin-engined widebody aircraft	双发宽体飞机
twin-spool axial flow turbo-propeller	双轴轴流式涡轮-螺旋桨发动机
twin-spool compressor	双转子压气机
twin-spool turboshaft	双轴涡轮轴发动机
twisted contour	扭曲外形
two dimensional inlet	二维进气道
two dimensional nozzle	二维喷管
two phase combustion	两相燃烧
two-phase-flow	二相流
two-position nozzle	双位喷口
Tyne	苔茵发动机(罗罗公司,英国)

type certificate holder	型号合格证持有者
type certification	型号合格证
type certification team	型号适航证工作组
type design change classification	型号设计更改分类
type design change procedure	型号设计更改程序
type investigation programme	型号事故调查程序
type of bearing	轴承型号
type of demonstration	验证方式
type test	型号试车
type test engine	发动机型号试车
Typhoon fighter	欧洲台风战斗机
Typhoon（Eurofighter）	台风-欧洲战斗机［欧洲（英、德、意、西四国）联合研制］
typical ear response	典型的听觉反应

Udimet	一种镍基耐热合金
Udimet 720 Li discs	材料为 Udimet 720 Li 的镍基涡轮盘
Ufa Engine Industrial Association/UMPO JSC	（俄罗斯）乌法发动机工业联合体股份公司
UK Defence Procurement Agency	英国国防部采购局
UK Generally Accepted Accounting Principles	英国公认会计原则
ultimate bearing capacity	轴承极限能力
ultimate load	极限载荷
ultimate tensile strength	极限拉伸强度
ultra fan	超扇发动机（罗罗公司，英国）
ultra green	超级环保
ultra high bypass ratio	超高涵道比
ultra high frequency	超高频率
ultra high propulsive efficiency	（齿轮传动）超高推进效率
ultra high-bypass	超高涵道比
ultra long-range aircraft	超长航程飞机

ultra low noise engine	超低噪声发动机
ultra low NOx	超低氮氧化物排放
ultra-efficient engine technology	（美国国家航空航天局的）极高效发动机技术（计划）
ultra-longhaul routes	超远程航线
ultra-micro combustor	超微型燃烧室
ultrasonic automated scanning system	超声自动化扫描成像检测系统
ultrasonic imaging system	超声成像检测系统
ultrasonic inspection	超声波检测
ultrasonic shot peening	超声喷丸
ultraviolet aging test	紫外老化试验
ultra-violet radiation	紫外辐射
unauthorised change	未经授权的变更
unbalance detector	不平衡探测器
unbalance sensor	不平衡传感器
unbalanced power	不平衡功率
unburned hydrocarbons	未燃碳氢化合物
uncertainty analysis	不确定性分析
uncontrolled fire	不可控的火情
undedicated research and development	非专用研发
underbead	焊缝底部
undercarriage track	主轮距
under-equipping	装备不足
underexpansion	不完全膨胀
underfueling	供油不足
undershaft feeds	从内轴输入滑油
underwing installation	翼下吊装
underwing mount model	翼下吊装模型
underwing/rear fuselage commonality	翼下/机身安装的通用性
undetected material defects	未探测的材料缺陷
unducted fan	无涵道风扇
unducted fan engine	无涵道风扇发动机
undulation	波纹
unexposed side of engine	发动机的非暴露边

unfavorable balance of trade	贸易逆差
unfused tacks	未融化的焊料
unidirectional lock	单向锁
uniform axial velocity	均匀的轴向速度
uniform quality standard	统一的质量标准
uniformly tight	紧度均匀
uninstalled engine	未安装的发动机
uninstalled performance	非安装性能
uninterruptible power supply	不间断电源
uniquely shaped blade	特有形状叶片
unit conversion	单位转换
unit positioning	元件定位
unit price	单价
unit removal	部件拆卸
United Aircraft Corporation	(俄罗斯)联合飞机制造集团
United Airlines	美联航
United Engine Corporation	(俄罗斯)联合发动机制造集团
United Kingdom Accreditation Service	英国测量认证机构
United States Government	美国政府
United States Marine Corps	美国海军陆战队
United Technologies Aerospace Systems	联合技术公司航空航天系统
United Technologies Corporation	(美国)联合技术公司
United Technologies Research Center	联合技术公司研究中心
universal asynchronous receiver/transmitter	通用同步接收器/发射器
universal engine vibration monitor	通用发动机振动监测仪
universal flight information system	全球飞行信息系统
universal gas constant	通用气体常数
universal wireless backbone system	(空客)通用无线主干系统
universe fan facility adaptation	通用风扇设备适应性
universities and aerospace institutions	大学及航空航天研究所
University of Karlsruhe Institute of Thermal Turbomachinery	(德国)卡尔斯鲁厄大学热叶轮机械研究院
University of Stuttgart Institute of Aero Propulsion Systems	(德国)斯图加特大学航空推进系统研究所

unloading cavity	卸载腔
unmanned aerial vehicle	无人机
unmanned aerial vehicles	无人飞行器
unmanned combat armed rotor craft	(美国)无人战斗武装旋翼机
unmanned combat air vehicle	无人战斗机
unmanned extended	无人机技术能力扩展
unpredictable costs	不可预见费用
unscheduled depot level maintenance	非计划内场级维修
unscheduled engine removal rate	提前换发率
unscheduled maintenance	非定期维护
unscheduled repair or removal	非计划修理或换发
unserviceable	无用的
unsteady aero forces	不稳定气动力
unsteady flow interaction	不稳定气流的相互作用
unstructured reactive	无组织的反应
unusual imperfection	异常缺陷
unusual rattles	异响
update of the engine type design definition	发动机型号设计定义更新
updated engine model	更新的发动机模型
upgraded cabin environmental control system	已升级的座舱环境控制系统
upper atmosphere	高层大气
upper control limit	控制上限
upper efflux	高位喷流
upper specification limit	规格上限
upper stage engine technology	(美国)高级发动机技术项目
US Air Force	美国空军装备总部
US Air Force Research Laboratory	美国空军研究实验室
US Department of Defense	美国国防部
US Department of Energy	美国能源部
US Department of Transportation	美国交通运输部
usable fuel	可用燃油
usage monitoring	使用监视
used life	已用寿命
useful plasma for aerodynamic control	(欧盟)实用等离子体气动控制项目

utilise revert   利用回收料

utilization   利用

vaccum arc remelting   真空自耗电弧熔烁

vaccum arc remelting furnace   真空自耗电弧熔烁炉

vaccum arc remelting skull furnace   真空自耗电弧熔烁凝壳炉

Vacu-Blast   (英国)喷丸/喷砂设备制造商

vacuum arc welding   真空电弧焊

vacuum assisted resin transfer molding   真空辅助树脂模塑工艺

vacuum braging furnace   真空钎焊炉

vacuum diffusion welding   真空扩散焊

vacuum electronic device   真空电子器件

vacuum form tooling   真空成形制模

vacuum heat treatment   真空热处理

vacuum hydrogen treatment furnace   真空氢化处理炉

vacuum plasma spraying   真空等离子喷涂

vacuum self-consumable skull furnace   真空自耗凝壳炉

valence analysis   价态分析

validation of type certification   型号适航证

validation requirement   验证要求

validation strategy report   验证策略报告

validation test   验证试验

validations of supplemental type certificate   补充型号认可证

value added time   有效时间

value engineering   价值工程

value improvement through a virtual aeron-   (欧盟项目)利用虚拟仿真方法实现
    autical collaborative enterprise   航空企业增值

value-added tax   增值税

values for customers   客户价值

valve-operating mechanism   分气机构

valves   气门,阀

vane type pump	叶轮泵
vaneless space	无导叶空间
vane-type computer numerical control shot peening machine	叶轮式数控抛丸机
Vantage	优势-公务机(视觉喷气机公司,美国)
vapor core pump	汽心泵
vapor deposition	气相沉积
vaporise the ice/water	蒸发冰/水
vaporization rate	蒸发率
vaporizer	蒸发器
vaporizing tube	汽化管
vapour gutter	蒸发槽
vapour locking	蒸气堵塞
variable area bullet drive	(喷口)可变面积微型驱动
variable area bypass injector	可调面积涵道引射器
variable area jet exhaust nozzle	可变面积尾喷管
variable area propelling nozzle	可变面积推进喷管
variable bleed valve	可调放气阀
variable camber inlet guide	变弯度导流叶片
variable circle engine	变循环发动机
variable cycle	变循环
variable fan nozzle	可调风扇导向器
variable frequency generator	变频发电机
variable geometric control	可变几何形状控制
variable geometry inlet	可变形状进气道
variable geometry turbine	变几何形状涡轮
variable guide vane	可调导向器
variable inlet	可调进气口
variable inlet guide vane	可调进气导向器
variable intake	可调进气道
variable metering orifice	可调节流油空
variable nozzle	可调喷口
variable oil reduction valve	可变减油阀
variable orifice	可调孔

variable propelling nozzle	可调推进喷管
variable restrictor	可调限流器
variable speed constant frequency	等频变速
variable stator vane	可调导向叶片
variable stator vane actuating system	可调导向叶片作动系统
variable stator vane actuator	可调静叶传动杆
variable stator vane electronic control mal-scheduling	可调导向叶片电控不当
variable stator vane malscheduling	可调导向叶片动作不当
variable stator vane mechanical malscheduling	可调导向叶片机械动作不当
variable stream control engine	变流路控制发动机
variable vane regulator	可调静叶调节器
variable yaw mechanism	变侧滑角调节机构
variable-cycle engines	可变循环发动机
variable-density tunnel	变密度风洞
variable-reluctance speed probe	可变磁阻式转速探头
variable air volume actuating system	变流量作动系统
variable flow ducted rocket	变流量涵道火箭
variable geometry combustor	变几何燃烧室
variable pitch mechanism	变迎角调节机构
various triggering modes	多种触发方式
varnish	清漆
Vautour	秃鹰-战斗机(达索公司,法国)
vectoring in flight	飞行矢量
vectoring nozzle	矢量喷管
vehicle charging and potential	飞行器充电和电压
vehicle-borne	车载
velocity characteristics	速度特性
velocity coefficient	速度系数
velocity gradient	速度梯度
velocity head	速度头
velocity of rotation	转速
velocity profile	速度剖面

velocity ratio	速度比
velocity triangle	速度三角形
vented aircraft	通风的飞机
venting system	通风系统
Venturi	文丘里(人名)
Venturitube	(用来制造压差)文氏管
Vericor Power Systems Inc.	(MTU 子公司)Vericor 动力系统公司
verify correct process routing	确认正确的加工步骤
versatile affordable advanced turbine engine	(美国)负担得起的多用途先进涡轮发动机
versatile core plan	(F136)通用核心机计划
vertical audit	垂直评审
vertical lift and up & away operating modes	垂直起飞和平飞操纵模式
vertical integration	纵向集成
vertical situation display	(飞机)高度状况显示
vertical stabilisers combined with winglets	垂直安定面与翼梢小翼融合
vertical take-off and landing	垂直起降
vertical/short take-off and landing	垂直/短距起降
verticality/short take-off and landing aircraft	垂直/短距起降飞机
verticality/short take-off and landing engine	垂直/短距起落发动机
verticality/short take-off and landing powerplant	垂直/短距起降动力装置
very high bypass	极高涵道比
very large transport airplane	超大型运输飞机
very light jet	超轻型喷气机
very light rotorcraft	超轻型旋翼机
very long range	超远程
very persistent and very bio-accumulative	高持续性和高生物积累性
vibrate gloss	振动抛光
vibration analysis system	振动分析系统
vibration diagnostic device	振动诊断仪器
vibration endurance	振动持久性
vibration isolation	隔振
vibration modal test	振动模态试验

vibration modes	振动模态
vibration pickup	拾振器
vibration test system	振动试验设备
vibration transmitter	振动传感器
vibro-meter	振动测量仪(公司)
vibro-polishing	振动磨光
videoscope	视频内窥镜
view along cord line	弦线视图
view looking rearward	后视图
Vikers	维克斯-客机(维克斯-威尔斯利公司,英国)
Vikers Armstrongs Aircraft Ltd.	(英国)维克斯-阿姆斯特朗飞机公司
Vikers Wellesely	(英国)维克斯-威尔斯利公司
Viking	维京海盗-反潜机(洛克希德-马丁公司,美国)
Viper	威派尔发动机(罗罗公司,英国)
Virginia Polytechnic Institute	弗吉尼亚理工大学
virtual disassembly/assembly technique	虚拟拆卸/装配技术
virtual enterprise	虚拟企业
virtual enterprise resource planning	虚拟企业资源计划
virtual manufacturing	虚拟制造
virtual private network	虚拟专用网络
virtual product development	虚拟产品开发
virtual product management	虚拟产品开发管理
virtual reality	虚拟现实
viscoelasticity	黏弹性
visco-plastic constitutive model	黏塑性本构模型
viscosity	黏度
viscosity-pressure coefficient	黏度压力系数
viscous creep	黏性蠕变
viscous damping	黏滞阻尼
Vision	远景技术规划(罗罗)
Visionaire Jet	(美国)视觉喷气机公司
visual angle	视场角

visual flight rules	目视飞行规则
visual inspection	目视检查
visualisation	可视性
visualiser	显示器
vital supplier	关键供应商
volatility	挥发性
Volt	(电压单位)伏特
voltage direct current	直流电
volume resistivity	体积电阻率
volumetric heat intensity	容热强度
Volvo Aero Corporation	(瑞典)沃尔沃航空工业公司
vortex burning and mixing	涡流燃烧与混合
vortex dynamics	涡动力学
vortex flap	涡襟翼
vortex generator	旋涡发生器
Vulcan	火神-轰炸机(霍克-西德利公司,英国)

wake vortex	机尾涡流
wake vortex-encounter	(飞机)尾涡干扰
wall contouring	壁板外形
wall section	壁面
wall temperature measurement	壁温测量
wall temperature of flame tube	火焰筒壁温
wall temperature test	壁温试验
Wankle engine	温克发动机(无活塞旋转式)
Wankle rotary engine	温克旋转式发动机
warehouse	仓库
warehousing facility	仓库
warranty administration	保修管理
washer	垫圈

Waspaloy	一种涡轮盘用镍合金
water cooled bearing	水冷轴承
water evaporation	水的蒸发
water ingestion test	吸水试验
water injection	喷水
water methanal control unit	(高压比发动机)水与甲醛比例控制装置
water proof test	防水试验
water rig testing	水模试验
water separator	水分离器
water sublimation	水的升华
water tunnel test	水槽试验
water vapour	水蒸气
water visulization test	水流模拟试验
water-bomber	灭火飞机
water-cooled heat shield	水冷隔热屏
waterfall chart	(表示三维的)瀑布图
waterjets	喷水式推进器
Watt	瓦
wave rotor topping cycle technology	波转子增压循环技术
wax injection	压蜡
wax mold	蜡模
weak extinction	贫油熄火
weapon pylon	武器挂架
wear	磨损
wear protection coatings	防磨损涂层
wear rate analysis	磨损率分析
web-based training	网上培训
web-enabled program	网上程序
Weibull analysis	威布尔分析法
Weibull distribution	威布尔分布
Weibull life distribution	威布尔寿命分布
weight breakdown	重量分解
weight of fuel	燃油重量

weight off wheels	起落架空载
weight on board	装载重量
weight on wheels	起落架负重
weight penalty	重量的代价
weighted accumulation of emission	排放污染的加权累积
weighted valve	配重活门
weld width	焊缝宽度
welded assembly	焊接装配
welded in pair	成对焊接
Welding Institute	(英国)焊接研究所
welding wire	焊接条
Welland engines	维兰德发动机
western first tiers	西方的一级供应商
western-built	西方制造
Westinghouse Electric Company	(美国)西屋电气公司
wet air pump	湿式空气泵
wet stream	湿蒸汽
wet thrust	含加力的推力
wet-spinning of refractory ceramics	难熔陶瓷湿纺
wheel spin up	轮子打滑
wheel up landing	无起落架着陆
whirl velocity control	涡流速度控制
Whittle-type turbo-jet engine	惠特尔型的一涡轮喷气发动机
whole engine definer	整体发动机方案制定者
whole engine model	整机模型
whole life cost	全寿命成本
wholly-owned subsidiary	独资子公司
who's who of commercial airlines companies	著名商业航空公司
wide bandwidth	宽频带
wide-bodied plane	宽体飞机
widebody	宽体飞机
widebody fuselage	宽体机身
wide-cut fuels	宽馏分燃油
Williams International	(美国)威廉姆斯国际公司

windage	风阻
windage effect	风阻效应
windmill relight	风车再点火
windmill starting	风车起动
windmilling，wind milling	风车状态
wing anti-ice system	机翼防除冰系统
wing anti-icing	机翼防除冰
wing leading edge intake	机翼前缘进气道
wing span	翼展
wing to fuselage fairing	机翼与机身间的整流罩
wing-mounted pod installation	翼下吊舱式安装
wire cutting machine	线切割机床
wire mesh support	丝网支架
wires and harnesses	管线
withdrawal paths	拆装回路
wood pattern	木制型模
work breakdown structure	任务分解结构
work in progress	进展中的工作
work package	工作包
work package owner	工作包负责人
work substances	工质
work team	工作小组
working cycle	工作循环
working line	工作线
working line threat	工作线变化
working line total pressure ratio	工作线的总增压比
works process specifications	工作流程规范（罗罗）
world-class capability	世界水平能力
Worldliner	环球班机
worm pump	螺杆泵
worsening pollution	日益严重的污染
woven fiber inner face sheet	机织纤维内表面层
woven part	机织材料零件
woven wire air-wetted surface overlay	机织金属防潮涂层

| Wright-Petterson Airforce Base | （美国）莱特-帕特森空军基地 |
| wrought superalloy | 变形高温合金 |

Xiamen's Taikoo Aircraft Engineering Company	厦门太古飞机工程公司
Xian Aero-Engines Corporation	西航公司
Xian Rolls-Royce Aerocomponents	西安西罗航空部件有限公司
X-ray fluorescence	X 射线荧光分析
X-ray inspection	X 光检测
X-ray test for high pressure spool clearance optimization	利用 X 光探测优化高压转子叶尖间隙

yacht	游艇
Yakovlev Design Bureau	（俄罗斯）雅克飞机设计局
yard	码
year to date	（指从会计年度开始到报表日的期间）本年度迄今
yield strength	屈服强度
yoke and field coils assembly	轭和励磁组件
yttrium aluminium garnet laser	钇铝石榴石激光

Zephyr	"西风"太阳能动力无人机
zero adjustment	调零
zero emission jet	零排放喷气飞机
zero escapes policy	无疵品政策
zero incidence of detected non-compliance	没有发现不合格品
zero line agreement	无疵品供货协议

zero power range	零功率区
zero tolerance policy	无宽容政策
zero variation and defects	零偏离无缺陷
zirconia	（用于涂层）氧化锆
zone extractor ventilation duct	区域抽气机通风管
"Z"-ring cooling	"Z"形冷却环

缩 略 词

3BSD	3 bearing swivel duct	3 轴承旋转喷管
3BSM	3 bearing swivel module	3 轴承旋转单元
3DRTM	3D resin transfer molding	三维树脂传递模塑成型
3DRTM	3D woven resin transfer molding	三维编织树脂传递模塑成型
3DSS	3D sensing system	三维扫描测量仪

A&M	approve & maintain	批准与维护
A&P	airframe and powerplant	机身与动力装置
A&TE	assembly and test engineering	组装和试验工程
A/C	aircraft	飞机
A/T	auto throttle	自动油门
AA	assembly area	组装区
AACUS	autonomous aerial cargo utility system	(美国)自主无人空中货运服务系统
AADC	Allison Advanced Development Company	(美国)艾利逊先进技术开发公司
AAIA	Air Accidents Investigation Branch	(英国)航空事故调查局
AAL	American Association for Laboratory	美国实验室协会
AALA	American Association for Laboratory Accreditation	美国实验室鉴定协会
AAQG	American Aerospace Quality Group	美国航空航天质量协调组织
AAR	air-to-air refuelling	空中加油
AATE	advanced affordable turbine engine	(美国)先进经济可承受涡轮发动机计划
AAVM	advanced airborne vibration monitor	先进的航空振动监测仪
AAW	active aeroelastic wing	主动气动弹性机翼
ABC	advancing blade concept	(直升机)前行桨叶方案
ABC	activity based costing	作业成本法
ABCCC	airborne command and control centre	机载战场指挥控制中心
AC	advisory circular	咨询通告
AC	airworthiness circular	适航通告

AC	alternating current	交流电
ACARE	Advisory Council for Aeronautic Research in Europe	欧洲航空研究咨询委员会
ACARS	aircraft communications addressing and reporting system	机载通信寻址和报告系统
ACAS	airborne collision avoidance system	机载防撞系统
ACC	active clearance control	主动间隙控制
ACC	active combustion control	主动燃烧控制
ACCA	advanced composite cargo aircraft	先进复合材料货运飞机
ACCEL	accelerating the electrification of flight	快速电动飞机项目(罗罗)
ACCESS	alternative fuel effects on contrails and cruise emissions	替代燃料对航迹和巡航排放的影响
ACE	adaptive cycle engine	(美国)适应性循环发动机
ACE	allied concurrent engineering	联合并行工程
ACEE	aircraft energy efficiency	飞机能耗效率
ACG	aviation capital group	航空金融集团
AC-HVAF	activated combustion high-velocity Air-Fuel Spraying	活性燃烧高速燃气喷涂
ACJ	Airbus Corporate Jetliner	空客公务喷气机
ACMG	Air Gargo Management Group	世界航空货运管理集团
ACMS	automated component mode synthesis	自动部件模态综合法
ACO	airspace control order	空中管制命令
ACOC	air cooled oil cooler	气冷滑油冷却器
ACS	accessory control specification	附件控制规范
ACS	aerial common sensor	空中通用传感器(美国侦察机)
ACS	automated commercial system	自动商务系统
ACSI	American Customer Satisfaction Index	美国客户满意指数
ACTIVE	Advanced Control Technology for Integrated Vehicles	一体化飞行器先进控制技术
ACU	acceleration control unit	加速控制装置
ACV	air control value	气控阀
ACX	advanced combat experimental	技术验证机(达索)
AD	airworthiness directive	适航指令

ADAM	Advanced Aeroengine Materials	（英国）先进航空发动机材料项目
ADC	air data computer	大气数据采集计算机
ADCC	Aviation Data Communication Corp.	中国民航总局航空数据通信公司
ADD	aircraft deferred defects	飞机延误故障
ADD	anti-dumping duty	反倾销税
ADECS	Adaptive Engine Control System	变循环发动机控制系统
ADF	automatic direction finder	自动定向仪
ADGT	aero-derivative gas turbine	航改型燃气涡轮机
ADI	advanced drawing issue	先期出图
ADP	advanced ducted propulsion	先进涵道推进
ADP	advanced ducted propulsor	先进涵道推进器
ADP	advanced demonstrator program	技术验证项目
ADSOV	active damper shut off valve	主动阻尼截止阀
Adv	advanced	先进的
ADVENT	Adaptive Versatile Engine Technology	（美国）自适应通用发动机技术项目
AE	Allison Engine Company（now Rolls-Royce Corporation）	艾力逊发动机公司（现为罗罗北美公司）
AEA	all electric aircraft	全电化飞机
AEA	Association of European Airlines	欧洲航空公司协会
AEC	Atomic Energy Commission	原子能委员会
AECMA	European Association of Aerospace Industries	欧洲航空航天工业协会
AECU	audio electronic control unit	音频电子控制单元
AEDC	Arnold Engineering Development Centre	（美国）阿诺德工程开发中心
AEE	All Electric Engine	全电化发动机
AEJPT	advanced european jet pilot training	先进欧洲喷气机驾驶员培训
AEM	advanced engineering memorandum	先进工程备忘录
AERL	aircraft engine research laboratory	飞机发动机研究实验室
AES	auger electron spectroscopy	俄歇电子能谱
AETD	adaptive engine technology development	（美国）自适应发动机开发项目

AETP	adaptive engine transition program	（美国）自适应发动机转换项目
AEW	airborne early warning	空中预警
AEWACS	Airborne Early Warning & Control System	空中预警与控制系统
AFDX	avionics full duplex switched ethernet	航空电子全双工交换式以太网
AFMC	Air Force Materiel Command	美国空军装备总部
AFR	air fuel ratio	空气与燃油比
AFRL	US Air Force Research Laboratory	美国空军研究实验室
AFS	AutoFlight System	自动飞行系统
AGARD	Advisory Group for Aerospace Research and Development	航空研发咨询委员会
AGATE	advanced general aviation transportation experiments	先进通航运输试验
AGB	accessory gear box	附件齿轮箱
AGB	angle gear box	角齿轮箱
AGR	advanced gas reactor	先进燃气发生器
AGT	advanced gas turbine	先进燃气涡轮
AGV	automated guided vehicle	无人搬运车
AH	attack helicopter	攻击直升机
AHFM	alternate hight-frequency material	交变高频材料
AHM	airplane health management	机队实时健康管理
AIA	Aerospace Industries Association of America	美国航空航天工业协会
AIAA	American Institute of Aeronautics and Astronautics	美国航空宇航研究所
AICS	air inlet control system	进气道控制系统
AICU	air inlet control unit	进气道控制单元
AIDC	Aerospace Industrial Development Corp.	（台湾）航空航天工业发展集团
AIDP	aircraft integration definition plan	飞机一体化计划
AIDS	aircraft integrated data system	飞机综合数据系统
AIMS	Advanced Integrated Manufacturing System	先进组合式制造系统

AIR	aerospace industry reports	航空航天工业报告
AIT	accident telex	事故通报
AIW	accelerated improvement workshop	精益管理(波音)
AJNC	active jet noise control	喷气噪声主动控制
AJT	advanced jet trainer	先进喷气式教练机
ALCA	advanced light combat aircraft	(捷克)先进轻型攻击机
ALCM	Air-Launched Cruise Missile	空中发射巡航导弹
ALECSys	advanced low emissions combustion systems	先进低排放贫油燃烧系统项目(罗罗)
ALI	airworthiness limitation item	适航限制项目
Al-Li	aluminum-lithium alloy	铝锂合金
ALM	application of lifecycle management	软件生命周期管理
ALT	Altitude	高度
AM	airworthiness manual	适航手册
AMAD	aircraft mounted accessory drive	机上附件传动装置
AMB	active magnetic bearing	主动电磁轴承
AMC	acceptable means of compliance	适航性验证方法
AML	anti-monopoly law	反垄断法
AMM	aircraft maintenance manual	飞机维护手册
AMMS	aircraft maintenance management system	飞机维护管理系统
AMO	approved maintenance organisation	经授权的维修机构
AMOF	authorized military overhaul facilities	经授权的军工大修设施
AMRTT	advanced multi-role tanker transport	先进多功能空中加油机
AMS	aerospace material specifications	航空航天材料规范
AMS	American Military Standard	美国军队标准
AMSL	above mean sea level	高出平均海平面
AMSS	aircraft maintenance support services	飞机维修支持服务
AMST	Advanced Medium STOL Transport	(美国)先进中型垂直起降运输机
AMT	accelerated mission test	加速任务试车
ANAC	Agencia Nacional de Aviacao Civil	巴西国家民航局
ANC	active noise control	噪声主动控制
ANCD	automated net control device	网络自动控制器
ANSI	American National Standards Institute	美国国家标准协会

ANTLE	affordable near term low emissions	(欧洲)可承受的近期低排放发动机项目
AoA	angle of attack	攻角
AOC	aerodrome obstruction chart	机场障碍图
AOD	accept on deviation	(特例接受有超差的部件)特采
AOD	available on demand	按需供货
AOD	argon oxygen decarburization	氩氧脱碳
AOG	aircraft on ground	飞机停航待修
AOHE	air/oil heat exchanger	油气热交换器
AoL	angle of incidence	入射角
AOPS	auxiliary oil pressure sensor	辅助油压传感器
AOT	all operators telex	全体用户通报
AP	accounting period	会计年度
APAQG	Asia-Pacific Aerospace Quality Group	亚太航空航天质量协调组织
APET	advanced prop-fan engine technology	先进桨扇发动机技术
APF	aft pylon fairing	后挂架整流罩
APM	airplane performance monitoring	飞机性能监测
APQP	advanced product quality planning	先进产品质量计划
APS	aftermarket purchase services	售后采购服务
APS	aircraft prepared for service	待命飞机(不含燃料和载荷)
APSC	active performance seeking control	主动寻找并控制最佳性能状态
APTD	Aerospace Propulsion Technology Demonstrator	(中国)航空推进技术验证项目
APU	auxiliary power unit	辅助动力装置
AQIS	Australian Quarantine and Inspection Service	澳大利亚检验检疫局
AQL	acceptable quality level	合格质量标准
AQP	advanced quality planning	先进质量计划
AR	advanced replacement	先期更换
AR	aspect ratio	展弦比
AR	augmented reality	增强现实技术
AR&O	aero repair and overhaul	航空发动机维修和大修

ARALL	aramid-aluminium fiber-metal laminate	聚芳族酰胺纤维铝合金层板
ARC	airframe related components	机体结构部件
ARH	armed reconnaissance helicopter	武装侦察直升机
ARI	aviation restructure initiative	航空业重组建议
ARINC	aeronautical radio incorporated	一体化航空无线电
ARMD	aeronautics research mission directorate	航空科研规划局
ARP	aerospace recommended practices	航空航天操作规程建议
AS	aerospace standard(s)	航空航天标准
AS9100	aerospace quality systems	航空航天质量体系标准
ASA	aviation suppliers association	航空供应商协会
ASB	alert service bulletin	报警服务公告
ASCM	aviation sub-contracting manufacture	航空转包生产
ASD	Aerospace and Defence Industries Association of Europe	欧洲航空航天和防务工业协会
ASD	aeronautical systems division	航空系统部
ASDA	accelerate-stop distance available	中断起飞可用场长度
ASE	aircraft survivability equipment	飞机生存性设备
ASEE	American Society for Engineering Education	美国工程教育协会
ASI	AirSpeed Indicator	空速表
ASIC	applications specific integrated circuit	特殊集成电路应用
ASK	available seat kilometers	可用座位公里
ASM	air separation module	空气分离模块
ASME	American Society of Mechanical Engineers	美国机械工程师协会
ASN	Aviation Safety Network	(美国)航空安全网
ASNM	available seats on aircraft per nautical mile	机上座位数/海里
ASNT	American Society of Non-Destructive Testing	美国无损检测协会
ASO	automation system operation	自动化系统运行
ASPIRE	advanced supersonic propulsion and integration research	先进超声速推进与集成研究
ASQM	aerospace standard for quality management	航空航天质量管理
ASR	air surveillance and reconnaissance	机载监视与侦察

AST	advanced supersonic transport	先进亚声速运输机
ASTM	American Society for Testing and Materials	美国材料和试验协会
ASTOVL	assisted short take-off and vertical landing	利用助推器短距起飞和垂直着陆
ASU	altitude sensing unit	高度传感装置
ATA	air transport association	空运协会
ATA	American Transport Association	美国运输协会
ATC	automatic throttle control	自动油门控制
ATC	air traffic control	空中交通管制
ATCC	air traffic control committee	空中交通管制委员会
ATCC	active tip-clearance control	主动叶尖间隙控制
ATD	actual time of departure	实际离开时间
ATE	automatic test equipment	自动测试设备
ATE	avionics testing equipment	机载设备测试装置
ATEC	advanced turbine engine company	先进涡轮发动机公司（霍尼韦尔与普惠合资）
ATEGG	advanced turbine engine gas generator	先进涡轮发动机燃气发生器
ATF	advanced tactical fighter	先进战术战斗机
ATF	altitude test facility	高空台
ATFCM	air traffic flow and capacity management	空中交通流量与容量管理
ATFI	advanced technology fan integration	一体化先进技术风扇
ATFM	air traffic flow management	空中交通流量管理
ATGS	auxiliary turbine-generator set	辅助涡燃发电机组
ATI	aerospace technology institute	（英国）航空技术研究所
ATIO	aviation technology，integration，and operations	航空技术，集成与运营
ATLA	air transport licensing authority	空运牌照局
ATM	air traffic management	空中交通管理
ATO	air traffic organisation	空中交通组织
ATO	authorized to offer	获准销售
ATO	after take-off	起飞后
ATO	aborted take-off	中断起飞
ATP	acceptance test procedure	验收测试步骤

ATP	advanced turbo-prop	先进涡桨发动机
ATP	aircraft test procedure	机上实验程序
ATP	approved to proceed	项目启动
ATP	advanced turboprop project	先进涡桨项目
ATR	attained turn rate	瞬时转弯速率
ATS	acceptance (production) test schedule	产品验收测试计划
ATS	air turbine starter	空气涡轮起动机
ATSB	Australian Transport Safety Bureau	澳大利亚运输安全局
ATT	advanced transport technology	先进运输技术
ATTAP	advanced turbine technology applications	先进涡轮技术应用
AUW	all up weight	飞机总重
AVD	application for variance/deviation	变更申请
AVGAS	Aviation Gasoline	航空汽油
AVIC I	China Aviation Industry Corporation I	中国航空工业第一集团公司
AVIC II	China Aviation Industry Corporation II	中国航空工业第二集团公司
AVICIT	AVIC Information Technology Co., Ltd.	金航数码科技公司
AVM	airborne vibration monitoring system	机载振动监视系统
Avro	A. V. Roe and Company	(英国)阿芙罗飞机公司
AVT	augmented vectored thrust	加力矢量推力
AVTUR	aviation turbine fuel	航空煤油
AVUM	aviation unit maintenance	航空设备维护
AWACS	airborne warning and control system	空中警告与控制系统
AWJ	abrasive waterjet	磨料水射流
AWS	American Welding Society	美国焊接协会
AWT	altitude wind tunnel	高空风洞
Ax	axial direction	轴向

B2B	business to business	企业对企业
B2C	business to customer	企业对消费者
BA	British Aircraft Corporation	英国飞机公司

BAE	British Aerospace Systems	（英国）英宇航公司
BAHE	buffer air heat exchanger	缓冲空气热交换器
BALP	Beijing Aero Lever Precision Limited	北京力威尔航空精密机械公司
BAM	buffer air manifold	缓冲空气总管
BAM	BeAnstandungsMeldung	疵品报告单
BAPS	buffer air pressure sensor	缓冲空气压力传感器
BASA	bilateral airworthiness and safety agreement	双边适航和安全协议
BAV	buffer air valve	缓冲气阀
BCAM	best cruise altitude and mach number	最佳巡航高度和马赫数
BCAR	British Civil Airworthiness Requirements	英国民用航空适航规定
BCCA	Bahrain Civil Aviation Affairs	巴林民航局
BCP	business continuity plan	业务持续计划
BDC	bottom dead centre	（动力）下止点
BDD	basic design data	基础设计数据
BDD	basic design document	基础设计文件
BDLI	Bundesverband der Deutschen Luft und RaumfahrtIndustrie E. V.	德国航空宇航工业联合会
BDS	balanced delivery score	平衡交货计分
BED	Bristol Engine Division	（英国）布里斯托尔发动机部
BER	beyond economic repair	不值得的维修
BER	business evaluation review	商务评审会
BERI	business environment risk intelligence	商务环境风险情报
BF	block fuel	航程总油耗（轮挡油耗）
BFCU	barometric flow control unit	气压式流量控制器
BFE	buyer furnished equipment	买方供货设备
BHP	brake horse power	制动马力
BIQ	build in quality	无缺陷
BITE	built-in test equipment	内置试验设备
BLING	bladed ring	整体叶环
BLISK	bladed disc	整体叶盘
BLUH	battlefield light utility helicopter	战场轻型通用直升机
BOAS	blade outer air seal	叶片外部空气封严
BOFM	bought-out finished materials	成品采购

BOM	bill of material	物料清单
BOP	bought-out parts	外购零件
BOT	burner outlet temperature	燃烧室出口温度
BPC	blade process casting	铸造叶片流程
BPF	blade passing frequency	叶片通过频率
BPF	blade process forging	锻造叶片流程
BPM	blade process machining	叶片机加工流程
BPM	business process model	业务流程模型
BPR	business process restructuring	重建业务流程
BPR	bypass ratio	涵道比
BPSD	Boeing Propulsion Systems Division	波音公司推进系统部
BPT	back pressure turbine	反压涡轮
BQS	balanced quality score	平衡质量记分
BR	book of reference	参考书
BRD	business requirements document	商务要求文件
BRI	bladed rotor integrity	叶片和转子一体化
BRJ	Bombardier regional jet	庞巴迪支线喷气机
BRR	BMW Rolls-Royce GmbH	(德国)宝马-罗罗公司
BRT	burner rig test	台架燃气模拟
BRW	break release weight	(飞机)最大允许起飞重量
BS	British Standard(s)	英国标准
BSBV	booster stage bleed valve	增压器级间放气阀
BSBV	buffer stage bleed valve	缓冲级间气阀
BSC	balanced score card	平衡记分卡
BSC	bunker surcharge	燃油附加费
BSEL	Bristol Siddeley Engines Ltd. (now Rolls-Royce)	布里托尔赛德利发动机有限公司(现属罗罗公司)
BSI	British Standards Institution	英国标准化协会
BSS	British Standard Specification	英国标准化规范
BSV	burner staging valve	分级燃烧阀
BT	block time	轮档时间(总航时)
BTO	build to order	按单订制
BTS	Bureau of Transportation Statistics	(美国)运输统计局
BTT	blade tip timing	叶尖定时测量

BTTF	blade tip timing frequency	叶尖定时频率
BTU	british thermal unit	英国热量单位
BV	bleed valve	放气阀
BVCU	bleed valve control unit	放气阀控制器
BVR	beyond visual range	超视距
BWB	blended wing body	翼身融合飞机

CAA	Civil Aviation Authority	英国民航局
CAAC	Civil Aviation Administration of China	中国民航局(原名)
CAAC	General Administration of Civil Aviation of China	中国民航总局(现名)
CAAM	Continued Airworthiness Assessment Methodologies	持续适航评估方法
CAAS	Civil Aviation Authority of Singapore	新加坡民航局
CAB	civil aeronautics board	(美国)民用航空委员会
CAD	computer aided design	计算机辅助设计
CADDS	Construction and Detailing Design System	构型和详细设计系统(设计软件)
CAE	China Aerospace Establishment	中国航空研究院
CAE	computer aided engineering	计算机辅助工程
CAEP	committee on aviation environmental protection	航空环境保护委员会
CAEW	compact airborne early warning	紧凑型空中预警飞机
CAF	currency adjustment factor	汇率调节附加费
CAGE	commercial and government entity	商业和政府权益
CAI	computer aided innovation	计算机辅助创新
CAI	contract assurance instruction	合同保证规定
CAIMS	central aircraft information maintenance system	飞机信息维护系统中心
CALS	commerce at light speed	光速商务
CAM	computer aided manufacture	计算机辅助制造

CAM	cost account manager	成本核算经理
CAP	combat air patrol	空中战斗巡逻
CAP	component approval package	部件核准包
CAP	computer aided plan	计算机辅助计划
CAPP	computer-aided process planning	计算机辅助工艺规划
CAPPS	computer assisted passenger prescreening system	计算机辅助乘客预先筛检系统
CAQC	computer aided quality control	计算机辅助质量控制
CAR	Canadian Aviation Regulation(s)	加拿大航空规定
CAR	corrective action report	纠正措施报告
CARD	civil aviation research and development	民用航空研发
CARE	carbon fibre reinforced aluminium	碳纤维铝合金层板
CAS	calibrated air speed	校正飞行速度
CAS	crew alert system	机组警报系统
CAS	crew altering system	机组人员变更系统
CASA	Civil Aviation Safety Authority	(澳大利亚)民航安全局
CASA	Construcciones Aeronáuticas S. A	西班牙航空工程公司
CASGC	China Aviation Suppliers Imp. & Exp. Group	中航材进出口集团公司
CASM	cost per available seat mile	每座英里成本
CASS	control and audit of specification system	特定系统的控制和审核
CAT1	Category 1	第一类
CATIA	Computer Aided Three-dimensional Interactive Application	计算机辅助三维交互式设计软件
CAUC	Civil Aviation University of China	中国民航大学
CAWI	China Aviation Weekly Index	中国航空每周指数
CBA	cobalt based alloy	钴基合金
CBD	cash before delivery	交货前付款
CBM	condition based maintenance	视情维护
CBR	consolidate business review	重组业务审查
CBT	computer based training	利用计算机培训
CCDR	customer critical design review	有客户代表参加的关键设计评审
CCE	chief controls engineer	负责控制的总师

CCF	conformance control feature	一致性控制特征
CCF	customer configuration file	客户构型档案
CCL	commerce control list	（美国）商务管制清单
CCL	compliance check list	一致性检查清单
CCOC	combustion chamber outer casing	燃烧室外部机匣
CCP	company calibration procedure	公司校准流程
CCP	component care process	部件维护过程
CCPP	combined-cycle power plant	联合循环发电厂
CCR	change control review	变更控制审核
CCR	customer change request	客户变更要求
CCS	create customer solutions	为客户创造解决方案
CCW	component cost worksheet	部件成本工作单
CCW	counter clockwise	逆时针
CCW	Custer Channel Wing	卡斯特沟槽机翼
CD	coefficient of drag	阻力系数
CDA	concept demonstration aircraft	（F135）方案验证机
CDA	concept design and analysis	方案设计和分析
CDA	continuous descent approach	连续下降进近
CDE	chief design engineer	总设计师
CDFS	core-driven fan stage	核心机驱动风扇级
CDG	component definition group	部件定义组
CDM	clean development mechanism	（京都协议减排）清洁发展机制
CDM	core design and manufacture	核心机的设计和制造
CDN	common data network	通用数据网络
CDP	component development plan	部件开发计划
CDP	component development programme	部件开发规划
CDP	compressor dischange temperature	压气机出口温度
CDP	compressor delivery pressure	压气机出口压力
CDR	critical design review	关键性设计评审
CD-ROM	compact disk-read only memory	光盘读入内存系统
CDU	cockpit display unit	驾驶舱显示装置
CE	concurrent engineering	并行工程

CECE	common extensible cryogenic engine	（美国国家航空航天局）通用可扩展的低温发动机
CEMCAN	ceramic matrix composite analyzer	陶瓷基复合材料分析器
CEN	computer added engineering	计算机辅助工程
CEO	current engine option	选装原发动机方案
CEP	component engineering procedure	部件设计流程
CER	certified emission reduction	核证的减排量
CER	cost estimate relation	成本估算法
CES	core engine size	核心机尺寸
CET	combustor exit temperature	燃烧室出口温度
CET	common external tariff	统一对外海关税率
CF	cumulative failures	累计失效次数
CFAN	Component Fan	（GE 和斯奈克玛公司合资）复合材料风扇叶片公司
CFBU	customer facing business unit	面对客户的业务单位
CFC	carbon fibre composite	碳纤维复合材料
CFC	chlorofluorocarbon	氟氯碳化物
CFD	computational fluid dynamics	计算流体力学
CFI	cumulative fatigue index	累计疲劳指数
CFMU	central flow management unit	中央流量管理中心
CFR	cost and freight	成本加运费
CFRP	carbon fiber-reinforced plastic	碳纤维增强塑料
CFRP	carbon fiber-reinforced polymer	碳纤维增强聚合物
CFX	Computational Fluid Dynamics	（英国）流体动力学计算软件
CHP	combined heat & power	热电联产
CIA	Central Intelligence Agency	中情局
CIAM	Central Institute of Aviation Motors	（俄罗斯）中央航空发动机研究院
CIC	close in combat	近距作战
CIC	compressor intermediate case	压气机中介机匣
CIF	cost，insurance and freight	成本、保险加运费
CIMS	computer integrated manufacturing system	计算机集成制造系统
CIP	Carriage and Insurance Paid	已付运费加保险费

CIP	Component Improvement Program	（美国）发动机部件改进计划
CIS	commercial invoice shipping	商业货运发票
CIS	customer information system	客户信息系统
CJT	cold junction temperature	冷结合温度
CL	coefficient of lift	升力系数
CLA	condition lever angle	油门控制位置
CLEAN SKY	clean sky joint technology initiative	（欧盟）清洁天空联合技术计划
CLEEN	continuous lower energy, emissions and noise	持续低能耗,低排放,低噪声
CLV	customer lifetime value	客户生命期价值
C-MAPSS	commercial modular aero-propulsion system simulation	商用模块化航空推进系统模拟
CMC	ceramic matrix composite	陶瓷复合材料
CME	company materials engineering	公司的材料技术（罗罗）
CMM	component maintenance manual	部件维护手册
CMM	coordinate measuring machine	坐标测量仪
CMO	current market outlook	当前市场展望
CMP	customer management point	客户管理点
CMR	certification maintenance requirement	审定维护要求
CMRT	carcinogens, mutagens and reproductive toxics	致癌物、诱变剂和可再生毒气
CMS	configuration management system	构型管理系统
CMT	component manufacturing technique	部件制造技术
CMTR	certified material test report	经认证的材料试验报告
CNC	computer numerical control	数控
CNG	compressed natural gas	压缩天然气
CO	carbon monoxide	一氧化碳
CO_2	carbon dioxide	二氧化碳
COC	cash operation cost	现金运营成本
CoC	certificate of conformity	质量合格证
COD	cash on delivery	货到付款
CODAC	configuration definition and control	方案的定义和控制
CODAG	combined diesel engines and gas turbine	柴油机和燃气轮机联合循环

CODOG	combined diesel engines or gas turbine	柴油机或燃气轮机联合循环
CoG	centre of gravity	重心
COM	costs of manufacture	制造成本
COMB	crew operation manual bulletin	机组操纵手册通告
CONQ	cost of non-quality	质量问题的代价
const	constant	常数
conv	conventional	惯用的,常规的
COP	commissioned overcheck policy	请第三方复检政策
COP	costs of production	生产成本
COPE	control pressure ratio engine	可控压比发动机
COS	condition of supply	供货条件
COSHH	control of substances hazardous to health（UK）	（英国）有害物质控制
COSS	combined overspeed & shut-off solenoid	组合超速限制螺线管
COT	combustor outlet temperature	燃烧室出口温度
CP	component proving	部件审核过程
CP	specific heat	定压热容
CPC	critical propulsion components	关键推进系统部件
CPE	chief performance engineer	性能总师
CPP	controllable pitch propeller	可控桨距螺旋桨
CPS	cycles per second	每秒循环数
CPU	central processing unit	中央处理器
CQAS	company quality acceptance standard	质量验收标准（罗罗）
CQCP	company quality control procedure	质量控制程序（罗罗）
CR	change request	变更要求
CR	corporate responsibility	企业社会职责
CR	concept review	方案审核
CRAIC	China-Russia Commercial Aircraft International Co.	中俄国际商用飞机公司
CRB	certification/registration bodies	认证注册机构
CRB	contract review board	合同审核委员会
CRI	certified release inspector	有资质的发货监察员
CRISP	counter rotating integrated shrouded propfan	（德国 MTU 公司）对转整体涵道式桨扇发动机

CRJ	Canada regional jet	庞巴迪支线客机(庞巴迪公司,加拿大)
CRM	common raw material	通用原材料
CRM	customer experience management	客户体验管理
CRM	customer relation management	客户关系管理
CROR	count rotating open rotor	对转桨扇开式转子发动机
CRP	controllable reversible pitch	可控逆桨距
CRP	contra rotating propeller	对转螺旋桨
CRR	certification readiness review	适航审核
CRT	cathode ray tube	阴极射线管
CSA	contractual service agreement	合同服务协议
CSAID	control system aircraft interface document	控制系统飞机界面文件
CSAM	contingency special airlift mission	(美国)紧急特殊空运任务
CSAR	combat search and rescue	战场搜寻和救援
CSC	customer support center	客户支援中心
CSCW	computer supported cooperative work	计算机协同工作
CSD	constant speed drive	恒速驱动
CSL	customer service and logistics	产品服务与支援
CSL	customized solutions	客户化方案
CSR	chemical safety report	化学品安全性报告
CSR	corporate social responsibility	企业社会职责
CSR	critical supplier review	重要供应商审核
CSS	commercial supply specification	商务供货规范
CSU	constant speed unit	恒速装置
CTA	Centro Tecnico Aeroespacial	巴西国家适航局
CTOL	conventional take-off and landing	常规起飞着陆
CTP	costed technical programme	项目工程成本
CTQ	critical to quality	品质关键
CV	control valve	控制阀
CV	carrier version	舰载型
CVC	constant volume combustion	等容燃烧
CVCCE	constant volume combustion cycle engine	等容燃烧循环发动机
CVD	countervailing duties	反补贴税
CVD	chemical vapor deposition	化学气相沉积

CVD	coating vapor deposition	气相沉积涂层
CVE	chief validation engineer	负责验证的总师
CVM	comparative vacuum monitoring	(无损探伤)比较真空监测
CVR	cabin voice recorder	座舱通话记录仪
CVR	cockpit voice recorder	驾驶舱录音机
CW	clock wise	顺时针
CW	compression work	压气机功
CWC	on-line customer web center	在线客户网络支持
CWO	cash with order	订货时付款
CWP	central warning panel	中央警告板
cwt	hundredweight	英担

D&M	design and make	设计并制造
D&C	delay and cancellation	延误和取消
D/C	date code	生产日期码
DA	decision altitude	决断高度
DA	defence aerospace	防务航空业务
DAB	pentagon's defense acquisition board	(美国)五角大楼国防采购部
DAC	data acquisition computer	数据采集计算机
DAC	double annular combustor	双环腔燃烧室
DAC	dual annular combustor	双环腔燃烧室
DAEM	deviation advanced engineering memorandum	先进工程超差备忘录
DAF	delivered at frontier	边境交货
DAR	drawing alteration request	图纸变更要求
DARPA	Defense Advanced Research Projects Agency	美国国防预先研究计划局
DAS	defensive aids suite	辅助防御系统
DAU	data acquisition unit	数据采集器
dB	decibel	(噪声单位)分贝
DB	diffusion bonded	扩散连接

DBAL	dynamic balance	动平衡
DBE	double bypass engine	双外涵道发动机
DBT	design-build teams	设计制造团队
DC	direct current	直流电
DCA	double circle arc	双圆弧形状
DCD	direct current dynamo	直流电机
DCL	declared cyclic life	呈报循环寿命
DCM	decontamination and cleaning methods	除污和净化方法
DCM	defence contract management	防务合同管理
DCMA	defence contract management agency	防务合同管理局
DCP	design change proposal	设计变更建议
DCU	data control unit	数据控制器
DCU	direction control unit	方向控制器
DCV	direction control valve	方向控制阀
DDIS	design definition issue statement	设计定义文件说明
DDIT	Demag Delaval Industrial Turbomachinery AB	西门子在瑞典的工业燃机分公司
DDP	declaration of design and performance	设计和性能的取证申报
DDP	delivered duty paid	完税后交货
DDR	deviation defect request	超差请求
DDRAM	design definition review and audit meeting	设计定义评审会
DDU	delivered duty unpaid	未完税交货
DECU	digital engine control unit	数字式发动机控制装置
DED	derby engine division	(英国)达比发动机部
DEEC	digital electronic engine control	数字式电子化发动机控制
DEFCON	defence condition	防务条件
DEFSPEC	defence specification（UK）	英国防务规范
DEFSTAN	defence standard（UK）	英国国防标准
DEFCS	digital electronic flight control system	数字电子飞控系统
DEG	degree	度数
Deg. C	degrees celsius	摄氏度(℃)
DEI	data element identification	数据元识别
DEL	demand export license	要求出口许可证

DELMIA	Digital Enterprise Lean Manufacturing Interactive Application	(法国达索)数字化企业互动精益制造软件
DEM	definitive engineering model	确定的工程模型
DEP	data entry plug	数据输入接口
DEQ	delivered ex quay	目的港码头交货
DER	designated engineering representative	(美国联邦航空局)委任工程代表
DES	delivered ex. ship	目的港船上交货
DESCP	develop existing supply chain process	现有供应链的开发过程
DFM	design for manufacture	可生产设计
DFM&A	design for manufacture & assembly	为制造和组装而设计
DFMEA	design failure mode effects analysis	设计故障模式和后果分析
DFSS	design for Six Sigma	六个希格玛设计
DFT	design for test	可测试设计
DG	diethylene glycol	二甘醇
DGA	French Defense Procurement Agency	法国国防采办局
DGAC	Direction Générale de l'Aviation Civile	法国民航总局
DGS	director general submarines	潜艇业务总经理
DH	decision height	决断高度
DHM	diagnostic health monitoring	诊断与状态监测
DHUD	digital head-up display	数字化平视显示器
DIA	detection/isolation/accommodation	故障检测隔离调整
Dia	diameter	直径
DIS	design issue standard	设计文件标准
DIS	drawing introduction sheet	图纸说明表
DKD	Deutscher Kalibrier Dienst (German Calibration Service)	德国标定服务中心
DLE	dry lean emission	干式低排放
DLE	dry low emission	干式低排放
DLEC	dry lean emission combustor	干式低排放燃烧室
DLO	defence logistics organisation	(英国)防务后勤机构
DLR	Deutsche zentrum fuer Luft-und Raumfahrt	德国航空航天研究院
DLR	German Aerospace Center	德国航空宇航研究中心
DM	data model	数模

DMAIC	define, measure, analyse, improve, control	定义、度量、分析、改进、控制
DMAP	Direct Matrix Abstraction Program	(NASTRAN 用户编程语言)直接矩阵提取编程
DMC	direct maintenance cost	直接维护成本
DMCG	direct maintenance cost guarantee	直接维护成本保证
DMDOR	designed modification design organisation representative	派驻改装设计机构的代表
DMSRR	development material specification	材料开发规范(罗罗)
DMU	digital mockup	数字样机
DNS	document numbering system	文件编号系统
DNW	Duits-Niederlande Windtunnels	德国荷兰风洞中心
DOA	design organisation approval	设计机构认证
DOC	direct operation cost	直接运营成本
DoD	US Department of Defense	(美国)国防部
DOE	design organization exposition	设计组织说明
DoE	US Department of Energy	美国能源部
DOH	design organization handbook	设计组织手册
DOS	door opening system	开门系统
DoT	US Department of Transportation	美国交通运输部
DOT	department of transportation	美国交通运输部
DP	design point	设计点
DPA	digital pre-assembly	数字化预装配
DPA	destructive physical analysis	破坏性物理分析
DPA	UK Defence Procurement Agency	英国国防部采购局
DPD	digital product definition	数字化产品设计
DPM	direct part marking	直接部件标记
DPU	defects per unit	缺陷率
DPWL	designated part warranted life	指定部件担保寿命
DQAG	defence quality assurance group	(英国国防部)防务质量保证组
DQR	designated quality representative	特派质量代表
DR	design review	设计评审
DRPC	design responsible party company	负责设计的合作公司

DRPS	development rationalised process specification	合理化开发流程规定
DRQSC	development rationalised quality standard component	合理化开发部件质量标准
DS	direct solidified	定向结晶
DS&S	data system and solution	数据系统和解决方案公司（罗罗）
DSCA	Defense Security Cooperation Agency	（美国）国防安全协作局
DSCL	declared safe cyclic life	取证申明安全循环寿命
DSIN	design scheme issue note	设计出图说明
DSM	drafting standards manual	标准化手册草案
DSP	digital signal processing	数字信号处理
DSU	data storage unite	数据存储器
DTC	data transfer cartridge	数据转换卡
DTC	Defence Technology Centre	国防技术中心
DTE	drains tank ejector	油箱排放喷射泵
DTED	digital terrain elevation data	数字化地形特征数据
DTI	Department of Trade and Industry, UK	英国贸工部
DTS	data transmission sheet	数据传输表
DTV	deflected thrust-vectoring	带倾角的推力矢量
DUTIFRISK	dual material titanium alloy friction welded blisk	（欧盟项目）双材料钛合金摩擦焊整体叶盘
DVR	design verification review	设计鉴定审核
DWT	dead weight tonnage	载重吨位

E3EEE	energy efficient engine	能源效率发动机
EADS	European Aeronautic Defence and Space Company	欧洲航空防务和航天公司
EAQG	European Aerospace Quality Group	欧洲航空航天质量组
EAR	Export Administration Regulations	（美国）出口行政规定

EARB	Export Administration Review Board	(美国)出口管理监管委员会
EAS	equivalent air speed	等效飞行速度
EASA	European Aviation Safety Agency	欧洲航空安全局(原JAA)
EASE	European Aerospace Supplier Evaluation	欧洲航空航天供应商评估
EAT	estimated approach time	预计进场时间
EB	electron beam	电子束
EBC	environmental barrier coating	环保型热障涂层
EBHA	electrical back-up hydrostatic actuator	液传作动器的电传备份
EB-PVD	electron beam-physical vaporation deposition	电子束物理气相沉积
EBU	engine build unit	组装好的发动机
EBU	engine business unit	发动机业务部
EBW	electron beam welding	电子束焊
EC	engineering change	技术变更
EC	European Commission	欧盟委员会
ECA	engineering control authority	工程控制主管机构
ECA	European Chemicals Agency	欧洲化学品管理局
ECAI	engine cowl anti-ice	发动机整流罩防冰
ECAM	electronic centralized aircraft monitor	飞机电子化集中监控器
ECCN	export control classification number	(美国)出口管制分类编码
ECCP	experimental clean combustor program	实验性低排放燃烧室计划
ECF	engine calibration facility	发动机标定设施
ECI	engine component improvement	发动机部件改善
ECIU	engine-cockpit interface unit	发动机与驾驶舱接口装置
ECM	electro chemical machining	电化学加工
ECM	electronic counter measures	电子对抗
ECM	engine control monitoring	发动机状态监视
ECN	engineering change notice	技术变更通知
ECP	engineering change package	技术变更包
ECPP	engineering critical parts plan	重要工程部件计划
ECR	engineering change request	技术变更请求
ECRD	engineering change release document	技术变更说明文件
ECS	environmental conditioning system	环境调节系统
ECS	environmental control system	环境控制系统

ECS	engine crank switch	发动机曲柄开关
ECU	electronic control unit	电子控制装置
EDC	engine data centre	发动机数据中心
EDE	enhanced durability engine	耐久性增强型发动机
EDI	engineering design instruction	工程设计指令
EDM	electronic data management	电子文件管理
EDM	engine design memorandum	发动机设计用文件
EDM	electro discharge machine	电火花加工
EDP	engine development plan	发动机开发计划
EDP	engine driven pump	发动机驱动泵
EDP	engine development programme	发动机开发规划
EDS	electronic data systems	电子数据系统
EDS	engineering design specification	工程设计规范（罗罗）
EDT	explanation of drawing terms	制图术语解释
EDU	engine display unit	发动机显示系统
EEC	engine electronic control	发动机电子控制器
EEC	European Economic Community	欧洲经济共同体
EEP	engine enhanced package	技术升级型发动机
EFC	engine flight cycle	发动机飞行循环
EFE	environmentally friendly engine program	（欧盟）环保发动机项目
EFH	engine flight hour	发动机飞行小时
EFIS	electronic flight information system	飞行仪表显示系统
EFPMS	electronic fuel pumping and metering system	燃油泵送和计量电子系统
EFQM	European Foundation for Quality Management	欧洲质量管理基金会
EFSEFD	endothermically fueled scramjet engine flight demonstrator	（美国）吸热超燃冲压发动机飞行验证机
EFT	estimated flight time	预计飞行时间
EFTA	European Free Trade Association	欧洲自由贸易联盟
EGA	electronic general arrangement	电子化总体结构图
EGT	exhaust gas temperature	排气温度
EGV	exite guide vane	出口导向叶片
EHA	electro hydrostatic actuator	电子液压作动器

EHD	elasto hydro-dynamic	弹性流体动力学
EHEST	European Helicopter Safety Team	欧洲直升机安全执行小组
EHM	engine health monitoring	发动机状态监视管理
EHSV	electro hydraulic servo valve	电液压伺服阀
EICAS	engine indicate crew alarm system	发动机指示机组报警系统
EICAS	engine indication and caution advisory system	发动机指示与警告提示系统
EIS	engineering inspection specification	工程检查规范
EIS	entry into service	投入运营
EIU	engine indicating unit	发动机显示器
EIU	engine interface unit	发动机接口元件
EKIP	Russian acronym: Ecology and Progress	埃基普(俄罗斯飞碟式航空器)
ELCS	engine loop control system	发动机环控系统
ELE	extended lean enterprise	扩展的精益企业
ELINT	Electronics Intelligence	电子情报
EM	engine manual	发动机手册
EMA	electro mechanical actuator	机电作动器
EMAS	Eco-management and audit scheme	(欧洲)生态管理和评估体系
EMC	electro magnetic compatibility	电磁兼容性
EMD	Engineering and Manufacturing Development	技术与制造开发
EMF	electro motive force	电动力
EMI	electro magnetic interference	电磁干扰
EMM	engine maintenance manual	发动机维护手册
EMP	engine management programme	发动机运营管理计划
EMS	electro magnetic susceptibility	电磁敏感性
EMS	engineering material specification	工程材料规范
EMS	environmental management system	环境管理系统
EMU	engine monitoring unit	发动机监控装置
ENFICA-FC	environmentally friendly inter city aircraft powered by fuel cells	(欧盟)燃料电池动力环保城际飞机项目
ENSIP	engine structural integrity program	发动机结构完整性计划
EOI	electro operating instructions	电动操纵说明书

EOI	engine operation instructions	发动机操纵手册
EOSID	engine out standard instrument departure	单发失效时标准仪表离场程序
EPA	Environmental Protection Agency	环保局
EPC	engineering, procurement and construction	设计、采购和施工
EPE	enhanced performance engine	性能提升型发动机
EPGDS	electrical power generation and distribution systems	发电和供电系统
EPI	Europrop International GmbH	欧洲动力国际公司（负责 TP400M 涡桨发动机）
EPM	electronic protection measures	电子保护措施
EPM	enabling propulsion materials	高效推进材料
EPNdB	effective perceived noise level in decibels	有效感知噪声
EPNdB	effective perceived noise in decibels	有效感知噪声
EPNL	effective perceived noise level	有效感知噪声级
EPR	engine pressure ratio	发动机增压比
EPS	engineering process specification	技术流程规范
ER	expansion ratio	膨胀比
ER	extended range	延程
ERA	environmentally responsible aviation	绿色航空
ERAST	environmental research and aircraft sensor technology	环境研究与飞机传感器技术
ERB	Engineering Review Board	技术审核委员会
ERB	engine research building	发动机研究大楼
ERFS	extended range fuel system	延程燃料系统
ERJ	Embraer Regional Jet	安博威支线客机（安博威公司,巴西）
ERM	elastomeric rubber materials	弹性橡胶材料
ERM	enterprise resource management	企业资源管理
ERMP	emergency response management plan	紧急响应管理计划
EROPS	extended range operations	延程运营
ERP	enterprise resource planning	企业资源计划
ERS	enhanced reference system	空间电子坐标系
ERT	emergency response team	应急响应小组

ESAD	equivalent still air distance	等效静风航程
ESAR	engineering source approval required	工程资源审批要求
ESAVD	electrostatic spray assisted vapor deposition	静电喷涂辅助沉积
ESC	engine supervisory control	发动机管理控制
ESD	engineering standards department	技术标准部
ESI	enterprise system integration	企业系统集成
ESM	electronic support measures	电子支持措施
ESP	engineering specification plan	技术要求计划
E-SR	EU standardisation report	欧盟标准化报告
ESS	engine section stator	一级导叶
ESS	engineering standards specification	工程标准规范
ESSD	electro static sensitive devices	静电敏感装置
ESSP	efficient small scale propulsion	高效小尺寸推进
ESTOL	extreme short takeoff and landing	超短距起降
ET&D	education training & development	教育培训和发展
ETAF	energy trends and alternate fuels	能源发展趋势与替代燃料
ETD	estimated time of departure	预计离港时间
ETE	estimated time en route	预计航线飞行时间
ETOPS	extended range twin operations	双发延程飞行
ETR	engineering technical report	技术支持报告
ETRAS	electrical thrust reverser actuation system	电作动反推系统(赛风)
ETS	emissions trading system	温室气体排放交易体系
EU	European Union	欧盟
EU FP	European Union Framework Program	欧盟框架科研计划
EVM	engine vibration monitoring	发动机振动监控
EVMU	engine vibration monitor unit	发动机振动监测仪
EVNRC	engine validation of noise reduction concepts	发动机降噪方案验证
EW	electronic warfare	电子战
EXW	Ex works	从工厂交货

| F | thrust | 推力 |

f	force	力
FAA	Federal Aviation Administration	美国联邦航空局
FADAS	fully automated data acquisition system	全自动数据采集系统
FadeA	Fabrica Argentina de Aviones	(阿根廷)阿根廷飞机制造厂
FADEC	full authority digital engine control	全权限数字发动机控制
FADEC	full authority digital electronic control	全权限数字电子控制
FAHP	fuzzy analytic hierarchy process	模糊层次分析法
FAI	first article inspection	首件产品检验
FAIR	first article inspection report	首件产品检验报告
FAPU	fuel cell based auxiliary power unit	燃料电池辅助动力装置
FAQ	frequently asked questions	常提出的问题
FAR	Federal Aviation Regulations	美国联邦航空条例
FAR	fuel/air ratio	油气比
FAS	free alongside ship	船边交货
FASCO	Shanghai Foreign Aviation Service Corporation	上海外航服务公司
FAST	Sichuan ChengFa Aero Science & Technology	成发科技公司
FATE	future affordable turbine engine	(美国)未来经济可承受涡轮发动机项目
FAV	fan air valve	风扇气阀
FBC	front bearing chamber	前轴承腔
FBI	Federal Bureau of Investigation	美国联邦调查局
FBL	fly-by-light	光传操纵系统
FBO	fan blade off	风扇叶片分离
FBO	fixed-base operation	固定运营基地
FBW	fly-by-wire	电传操纵系统
FC	flight cycle	飞行循环
FCA	free carrier	承运人交货
FCBA	future carrier-borne aircraft	(英国)未来舰载机
FCC	flight control computer	飞控计算机
FCM	flight certification modification	飞行审定变更
FCOC	fuel cooled oil cooler	用燃油冷却的滑油冷却器
FCOM	flight crew operating manual	机组人员操作手册

FCPA	Foreign Corrupt Practices Act（USA）	（美国）海外反腐败法
FCRC	fuel consumption retention guarantee	耗油率衰减保证
FCU	fuel control unit	燃油控制装置
FDGS	fan drive gear system	风扇驱动齿轮系统
FDM	fused deposition modeling	熔融沉积成型工艺
FDR	flight data recorder	飞行数据记录仪（黑匣子）
FDSS	fault detection subsystem	故障探测子系统
FDV	flow divider valve	分流阀
FEA	finite element analysis	有限元分析
FEGV	fan exit guide vane	风扇出口导向器
FEM	finite element method	有限元法
FET	fighter engine team	战斗机发动机工作组
FETT	first engine test team	首次发动机试车工作组
FFDP	fuel filter differential pressure sensor	燃油过滤器压差传感器
FFET	first full-engine test	首次整机试车
FFG	fuel flow governor	燃油油量调节器
FFR	fuel flow regulator	燃油流量调节器
FFT	fast fourier transform	快速傅里叶变换
FGD	fuel gas desulphurization	燃气脱硫
FGI	finished goods inventory	成品存货
FHA	fleet hour agreements	按机队飞行小时付费协议
FI	failure intensity	失效密度
FIB	focused ion beam	聚焦电子束
FIC	fan intermediate case	风扇中介机匣
FICA	fault indication and correction action	故障指示和排除
FICD	functional interface control document	界面控制功能文件
FIFO	first in first out	先进先出
FIM	fault isolation manual	故障隔离手册
FIPS	Federal Information Processing Standards（USA）	（美国）联邦信息处理标准
FJCA	future joint combat aircraft（JSF）	（美国）未来联合战斗机
FJGT	flame jet gradient test	火焰喷射梯度试验
FL	flight level	标准气压高度层
FLA	flight hour agreement	保修协议（GE）

FLADE	fan-on-blade	叶端风扇
FMC	flutter mode control	颤振模态控制
FMEA	failure modes and effects analysis	故障模式和后果分析
FMEA	frequent malfunction emergency analysis	常见故障紧急分析
FML	fiber-metal laminates	纤维金属层压板
FMMDS	forging method of manufacture data sheets	生产数据表中的锻造法
FMO	fuel metering orifice	燃油表气口
FMOD	flight modification approval	飞行变更批准
FMS	flexible manufacturing system	柔性制造系统
FMS	flight management system	飞机管理系统
FMS	flight meteorological service	飞行气象服务
FMS	fuel management system	燃油管理系统
FMU	fuel metering unit	燃油表
FMV	fuel metering valve	燃油表阀门
FMV	Swedish Defence Material Administration	瑞典国防装备管理局
FNF	fault not found	未发现故障
FOAS	future offensive air system	(英国)未来空中攻击系统
FOB	free on board	船上交货
FOC	free of charge	免费
FOC	full operational capability	具有全面运营能力
FOD	foreign object damage	外来物损伤
FOD	foreign object debris	异物碎屑
FOHE	fuel/oil heat exchanger	用燃油冷却滑油的热交换器
FOHEBV	fuel/oil heat exchanger bypass valve	燃油-滑油热交换器涵道阀门
FOQA	flight operational quality assurance	飞行运营质量保证
FOT	flight operations telex	飞行运营通报
FOT&E	follow-on test and evaluation	继续测试和评估
FP	feathering pump	顺桨泵
FP	fuel pump	燃油泵
FP7	EU's Seventh Framework Program	欧盟第七框架科研计划
FPD	fixed process document	生产流程管控文件
FPI	fluorescent penetrant inspection	荧光渗透检测

FPP	fixed pitch propeller	定距螺旋桨
FPR	faulty part report	报废零件报告
FPS	feet per second	每秒英尺
FPS	finished part stores	成品零件库
FPS	flight propulsion system	飞行推进系统
FPSOV	floating production, storage and offloading vessel	浮式石油生产储存装卸船
FPSU	Floating Production Storage Unit	浮动生产储油船
FPU	Floating Production Unit	浮动采油船
FRACAS	failure reporting, analysis & corrective actions systems	故障报告分析和排除系统
FRCI	fibrous refractory composite insulation	(美国国家航空航天局)碳纤维复合绝热材料
FRP	fiber reinforced plastics	纤维加强塑料
FRP	framework program	(欧盟科研)框架项目
FRR	flight readiness review	待飞审查
FRSI	felt reusable surface insulation	(美国国家航空航天局)柔性可重复使用表面绝热材料
FRV	fuel return valve	燃油回流阀
FSB	Flight Standardization Board	飞行标准化委员会
FSC	front split casing	前部整流机匣
FSER	full-scale engine research	全尺寸发动机研究
FSN	fuel spray nozzle	喷油嘴
FSS	flight service station	飞行服务站
FSTA	future strategic tanker aircraft	(英国)未来战略加油机
FSW	friction stir welding	搅拌摩擦焊
ft	foot	英尺
FTA	fault tree analysis	故障树分析法
FTA	free trade area	自由贸易区
FTB	flying test bed	飞行试车台
FTE	flight test engine	试飞用发动机
FTIS	fourier transform infrared spectroscopy	傅里叶红外光谱
FTIS	fuel tank inerting system	燃油箱惰化系统

FTK	freight tonne kilometers	货运吨公里
FTO	foreign trade organization	国际贸易组织
FTP	flight test programme	试飞项目
FUSCOMP	fuselage composite	(法国政府资助的)全复合材料机身计划
FW	fixed wing	固定翼
FWD	forward	向前
FWHR	feed water heater repowering	用供水型加热器增加功率

g	gram	克
GA	general arrangement	(发动机)总体结构图
GAE	grid-aided engineering	网格辅助工程
GAMA	General Aviation Manufacturers Association	通用航空生产商联盟
GAO	Government Accountability Office	(美国)政府审计署
GAP	general aviation propulsion	通用航空推进系统
GAP	Guggenheim Aviation Partners	(美国飞机租赁公司)古根海姆航空伙伴
GAPL	general arrangement parts list	总体结构零件目录
GATT	general agreement on tariffs and trade	关税与贸易总协定
GBAD	ground-based air defences	陆基防空系统
GC	general performance-control capability	总体绩效控制能力
GCE	global coordinating enviroment	全球协作环境
GCRS	global component repair services	全球部件维修服务
GDMS	glow discharge mass spectrometry	辉光放电质谱
GDP	gross domestic product	国内生产总值
GDT	ground development test	地面研发试验
GE	General Electric	通用电气公司
GEAE	GE Aircraft Engines	通用电气集团飞机发动机公司
GECAS	GE Capital Aviation Service	通用电气集团金融航空服务公司

GED	General Equipment Department of Chinese People's Liberation Army	中国人民解放军（原）总装备部
GEM	global engineering method	全球统一技术方法
GEnx	general electric next-generation	GE 为 B787 研制的发动机
GER	global engineering reference	全球技术参考资料
GEW	ground effect wing	气垫机翼
GFE	government funded equipment	政府资助设备
GHG	greenhouse gas	温室效应
GIS	green investment scheme	绿色投资计划
GKN	Guest Keen & Nettlefolds Aerospace	吉凯恩（客人-敏锐-内特尔福德）航宇公司
GLACIER	Global Aerospace Centre for Icing and Environmental Research	全球航空结冰与环境研究中心（罗罗）
GLARE	glass-reinforced fiber metal laminate	格拉尔叠层版（荷兰代尔芙特大学开发的玻璃纤维金属混合夹层板）
GM	green manufacturing	绿色制造
GMAW	gas metal arc welding	气体保护金属级电弧焊
GMS	global material solutions	（普惠公司专为 CFM56 生产备件的）全球化备件方案
GNP	gross national product	国民生产总值
GOCO	government owned contractor operated	国有民营的管理模式
GPA	general performance access	总体绩效评估
GPC	Global Procurement Council	全球采购理事会
GPD	global product development	全球产品研发模式
GPH	gallons per hour	每小时加仑
GPM	gallons per minute	每分钟加仑
GPQSB	Global Process Quality System Board	全球质量系统流程委员会
GPS	global positioning system	全球定位系统
GPU	ground power unit	地面电源
GQP	group quality procedure	团队质量程序
GR	goods receiving/received	收货
GRE	ground run-up enclosure	地面试车准备区
GRIP	goods receiving input process	收货登记流程

GS	general specification	一般规定
GS	global scorecard	全球记分卡
GSE	ground support equipment	地面支持设备
GSP	global supplier portal	全球供应商门户
GSS	ground support system	地面支援系统
GT	group technology	成组技术
GTE	China Gas Turbine Establishment	中国燃气涡轮研究院
GTF	geared turbo-fan	齿轮传动涡扇发动机
GTGS	gas turbine generator set	燃气涡转发电机组
GTM	Global Trade Management	国际贸易管理

H&S	health & safety	健康与安全
HA	horizontal audit	水平审计(同级别间审计)
HAESL	Hong Kong Aero Engine Services Limited	香港航空发动机服务公司
HAL	Hindustan Aeronautics Ltd.	印度锡都斯坦航空公司
HALE	high – altitude long – endurance	高空长航时
HARV	high alpha research vehicle	高攻角实验机(NASA)
HAS	high alloy steel	高质合金钢
HAT	height above touchdown	距着陆区高度
HATP	high-angle-of-attack technology program	高攻角技术项目
HB	Hardness-Brinell	(瑞典人 Brinell 定义的)布里涅尔硬度
HBV	handling bleed valve	控制放气阀
HC	hydrocarbons	碳氢化合物
HCF	high cycle fatigue	高周疲劳
HCU	hydraulic control unit	液压控制器
HDTO	hot day take off	高气温起飞
HEAT	high effectiveness advanced turbine	高效先进涡轮
HEDG	high efficiency deep grinding	高效深度磨削
HEETE	highly efficient embedded turbine engine	(美国)高效埋入式涡轮发动机

HEX	heat exchanger	热交换器
Hg	mercury	汞柱
HGS	head-up guidance system	(飞机着陆)平视导航系统
HHV	high-heating value	高热值
HIDEC	highly integrated digital electronic control	高度集成数字式电控
HIGE	hovering in ground effect	地面效应悬停
HiMaTE	high mach turbine engine	高马赫数涡轮发动机
HIRF	high intensity radiated fields	高强度辐射性场
HiSAC	environmentally friendly high-speed aircraft	(欧盟)环境友好型高速飞机项目
HISTEC	high stability engine control	高稳定性发动机控制
HiSTED	high-speed turbine engine demonstrator	(美国)高速涡轮发动机验证机(计划)
HITEMP	advanced high temperature engine materials technology program	先进发动机高温材料技术项目
HIVA	hazard identification and vulnerability assessment	危害确认和漏洞评估
HLFC	hybrid laminar-flow control	混合式层流控制
HMD	helmet mounted display	头盔显示器
HMP	hazard mitigation program	减灾计划
HMS	heat management system	热处理系统
HMU	hydro-mechanical unit	液压机械组件
HOGE	hovering out of ground effect	离地悬停
HOR	hold open rod	保持打开支撑杆
HP	high pressure	高压
HP	hydraulic pump	液压泵
hp	horsepower	马力
HPC	high pressure compressor	高压压气机
HPSOV	high pressure shut-off valve	高压节流阀
HPT	high pressure turbine	高压涡轮
HPV	high pressure valve	高压阀
HPV	high pressure vessel	高压容器
HQ	Herschel-Quincke tubes	赫歇尔-昆克管
HQO	higher quality officer	(英国国防部)高级质量官员

hr/hrs	hour/hours	小时
HRC	Rockwell C hardness	（罗克韦尔公司定义的）洛氏 C 级硬度
HRC	historical reference collection	历史性参考文献汇编
HRSG	heat recovery steam generator	热回收蒸汽发生器
HRSI	high-temperature reusable surface insulation	（美国国家航空航天局）高温重复使用表面绝热材料
HS&E	health safety & environment	健康安全与环境
HSCT	high speed civil transport	高速民用运输机
HSPT	high speed power turbine	高速动力涡轮
HSR	high speed research	（美国国家航空航天局）高速研究计划
HSR-EPM	high speed research-enabling propulsion materials	（美国国家航空航天局）高速研究计划－高效推进材料
HT	high tension	高张力
HTDU	high temperature demonstration unit	高温验证机
HTFC	high-temperature fuel cell	高温燃料电池
HUD	head-up display	平视显示器
HV	hardness-Vickers	（Vickers 定义的）威氏硬度
HVOF	high velocity oxygen fuel	高速火焰（喷涂）
HVT	high value target	高值目标
HYD	hydraulic	水力的
HySET	hydrocarbon scramjet engine technology	用碳氢化合物燃料的冲压发动机技术
Hz	Hertz	赫兹

I&S	investments and services	投资和服务
IAC	Interstate Aviation Committee	俄罗斯适航局
IACAR	Interstate Aviation Committee Aviation Register	俄罗斯适航局航空记录

IAE	International Aero Engines	国际航空发动机公司
IAI	Israel Aerospace Industries	以色列航空航天工业公司
IAP	integrated actuation package	综合电动液压作动器
IAQG	International Aerospace Quality Group	国际航空航天质量协会
IAS	indicated air speed	实测相对飞行速度
IATA	International Air Transport Association	国际航空运输协会
IBEM	inverse boundary element method	反向边界元法
IBR	integrally bladed ring	整体叶片环
IBR	integrally bladed rotor	整体叶片转子
IBV	interstage bleed valve	级间放气阀
IC	inspection certificate	检验证书
ICAO	International Civil Aviation Organization	国际民航组织
ICAOA	International Civil Aviation Organization Assembly	国际民航组织大会
ICAS	International Congress of Aeronautical Sciences	航空科学国际大会
ICD	interface control document	接口控制文件
ICE	internal combustion engine	内燃机
Icell	ice cell	冰室
ICGS	integrated coast guard system	联合海岸防卫系统
ICN	IATA Code Number	国际航空运输协会编号
ICR	intercooled recuperated	间冷回热
ICS	IAE Company Supplier	国际发动机公司的供应商
ICSS	integrated control and safety system	综合控制和安全系统
ICU	isolation control unit	隔离控制系统
ICV	isolation control valve	隔离控制阀
ID	inner diameter	内径
IDD	integration definition document	一体化方案文件
IDEEC	improved digital electronic engine controller	改进的数字电子式发动机控制器
IDF	Indigenous Defense Fighter	(台湾)经国号战斗机
IDG	integrated drive generator	组合驱动发电机
IDOC	indirect operating costs	间接运营成本
IDS	integrated design system	一体化设计系统

IED	integrated electric drive	整体式电驱动
IEDP	inertial engine development program	初始发动机研发项目
IEM	interpretive explanation material	(适航要求)说明性解释资料
IEMS	integrated engine management system	综合发动机管理系统
IETF	inverted exhaust turbofan	倒置排气涡扇发动机
IETM	interactive electronic technical manual	交互式电子技术手册
IFATS	innovative future air transport system	(欧盟)创新型未来空运系统
IFCS	intelligent flight control system	智能飞机控制系统
IFE	in-flight entertainment	飞行娱乐
IFO	International Field Office (FAA)	(美国联邦航空局)外航业务办公室
IFPC	integrated fuel pump and control	一体化燃油泵与控制系统
IFR	instrument flight rules	靠仪表盲飞规则
IFRS	international financial reporting standards	国际财务报告标准
IFS	inner fixed structure	内部固定结构
IFSD	in-flight shut down	空中停车
IFSDR	in-flight shut down rate	空中停车率
IFU	International Field Unit (FAA)	(美国联邦航空局)外航业务处
IFW	inertial friction welding	惯性摩擦焊
IGCC	integrated gasification combined-cycle	一体化的气化联合循环
IGV	inlet guide vane	进气导向叶片
IHI	Ishikawajima-Harima Heavy Industries Co. , Ltd	(日本)石川岛播磨重工业株式会社
IHPTET	integrated high performance turbine engine technology	(美国)高性能涡轮发动机综合技术项目
IHRC	inspection history record card	检验历史记录卡
ILAC	International Laboratory Accreditation Cooperation	国际实验室认证协作机构
ILFC	International Leasing and Finance Corporation	国际金融租赁公司
ILS	instrument landing system	仪表着陆系统
ILTV	inner loop thrust vectoring	内环推力矢量
IM	induction motor	感应电机

IM	installation memorandum	发动机安装备忘录
IMA	integrated modular avionics	综合模块式航电
IMC	intermetallic compound	金属间化合物
IMF	International Monetary Fund	国际货币基金组织
IMI	Israeli Military Industries	以色列军事工业公司
IMIS	integrated maintenance information system	综合维护信息系统
IMN	indicated mach number	实测马赫数
IMU	inertial measurement unit	惯性测量系统
in	inch	英寸
INS	inertial navigation system	惯行导航系统
IOP	independent overspeed protection	独立超速保护
IP	intermediate pressure	中压
IP3E	integrated product development, production and maintenance	一体化产品开发、生产和维护
IPC	illustrated parts catalogue	图示部件目录
IPC	inspection procedure card	检验流程卡
IPC	intermediate pressure compressor	中压压气机
IPCKV	inlet pressure check valve	进气压力检查阀
IPCS	indirect purchasing control specifications	间接采购控制规定
IPCS	integrated propulsion control system	一体化推进控制系统
IPLC	integrated propulsion, lift and control	动力、增升和控制一体化
IPMT	integrated project management team	综合项目管理团队
IPO	initial public offering	原始股上市
IPP	improving process performance	提高流程能力
IPPS	fully-integrated powerplant system	一体化推进系统
IPQC	in process quality control	制程中的质量管理
IPR	intellectual property rights	知识产权
IPS	inlet particle separator	进气道颗粒分离器
IPS	integrated propulsion system	一体化推进系统
IPT	integrated product team	综合产品开发组
IPT	integrated programme team	综合项目小组
IQC	incoming quality control	进料质量管理
IQN	invoice query notification	账单查询通知
IRA	intercooled recuperated aero-engine	间冷回热航空发动机

IRAC	integrated resilient aircraft control	一体化弹性飞机控制
IRC	Interdisciplinary Research Centre	跨学科研究中心
IRCA	international register of certificated auditors	有资质的审核员国际注册
IRCM	infrared countermeasures	红外对抗
IRD	integration requirements document	一体化要求文件
IRDS	infrared detection set	红外探测仪
IRR	internal rate of return	内部盈利率
IRST	infrared search and track	红外线搜索跟踪系统
ISA	International Standard Atmosphere	国际标准大气
ISD	in service date(s)	在役天数
ISFD	integrated standby flight display	一体化待机飞行显示
ISG	integrated starter and generator	一体化的启动机和发电机
ISIS	integrated standby instrument system	综合应急仪表系统(GE)
ISO	international standards organisation	国际标准化组织
ISO14001	the international standard for environmental management systems	国际标准化环境管理体系
ISO9001	the international standard for quality management systems	国际标准化质量管理体系
ISPM	international standard for phytosanitary measures	国际植物检疫标准
ISS	international space station	国际空间站
ISS	internet security system	互联网安全系统
IST	instrumentation standards and techniques	测试标准和技术
IT	information technology	信息技术
ITAR	International Traffic in Arms Regulations	(美国)武器国际贸易条例
ITB	inter turbine burning	(美国)内燃涡轮技术
ITEP	improved turbine engine program	(美国陆军)改进涡轮发动机计划
ITFU	intensive flying trials unit	(英国)强化试飞队
ITP	Industria de Turbo Propulsores	(西班牙)涡轮发动机工业公司
ITS	intelligent transportation system	智能交通
ITT	Inter-Turbine Temperature	级间涡轮温度

ITT	invitation to tender	招标
IVB	inner V blade	内 V 型叶片
IVCS	integrated voice communication system	飞行器通信系统
IVP	inverted velocity profile	倒置速度截面

J	Joule	焦耳
JAA	Joint Aviation Authorities	欧洲联合航空局
JAAT	Joint Aviation Authorities Transition	欧洲联合航空局过渡性机构
JAEA	Japan Aerospace Exploration Agency	日本航空航天开发机构
JAEC	Japanese Aero Engine Corporation	日本航空发动机公司
JANS	Joint Army-Navy Specifications（USA）	（美国）陆军和海军通用规范
JAR	joint airworthiness requirement	欧洲联合适航要求
JAR-E	joint aviation requirement-engine	欧洲联合适航对发动机的要求
JAR	joint aviation requirements	欧洲联合航空规范
JCA	joint cargo aircraft	（美国陆军和空军）联合运输机
JDB	joint definition bulletin	联合定义公告
JDP	joint definition phase	联合定义阶段
JDS	joint design standard	联合设计标准
JEP	joint engineering procedure	联合工程程序
JES	joint engineering specifications	联合工程规范
JES	joint engineering standard	联合工程标准
JETEC	joint expendable turbine engine concept	（美国）一次性涡轮发动机综合方案
JIT	just-in-time	准时
JMR	joint multi-role rotorcraft	联合多用途旋翼机
JOSV	journal oil shuttle valve	油梭往复阀
JPO	joint program office	综合计划办公室
JPT	jet pipe temperature	喷管温度
JSF	joint strike fighter（Lightning Ⅱ）	（美国）联合攻击机(闪电Ⅱ)

JTDE	joint technology demonstrator engine	综合技术验证发动机
JTI	Joint Technology Initiative	（欧盟项目）联合技术计划
JTIDS	joint tactical information distribution system	联合战术信息分布系统
JTM	joint technical meeting	联席技术会议
J-UCAV	joint-unmanned combat air vehicle	联合无人战斗机
JV	joint venture	合资
JVC	joint venture company	合资公司

K	absolute temperature Kelvin	绝对温度（K）
KAI	Korea Aerospace Industries Ltd.	韩国航空工业公司
KBE	knowledge based engineering	知识工程
KC	key characteristics	主要特性
kg	kilogram	千克
KHI	Kawasaki Heavy Industries	（日本）川崎重工公司
kJ	Kilojoule	千焦耳
KLIVT	klimov vectoring thrust	克里莫夫矢量推力
KM	kinetic metallization	动态金属涂层
km	kilometre	千米
kn	knot	节（速度）
kPa	kilopascal	千帕斯卡
KPI	key performance indicator	主要性能显示器
KSAPCA	Kingdom of Saudi Arabia Presidency of Civil Aviation	沙特阿拉伯王国民航局
kW	kilowatt	千瓦

L/C	letter of credit	信用证
L/D	length/diameter	长/径比
L/D	lift/drag ratio	升/阻比

L/N	lot number	批号
L/T	lead time	制造周期
LAIR	last article inspection report	末件产品检验报告
LAP	latch access panel	带有门闩的检修面板
LAP	large-scale advanced propfan	先进桨扇发动机大型验证机
LAS	low alloy steel	低级合金钢
Lasform	laser formation	激光成形
LAX	Los Angeles International Airport	洛杉矶国际机场
lb	pound	磅
LBA	Luftfahrt-Bundesamt	德国民航局
lbf	pounds force	磅力
lb-ft	pounds-foot（feet）	磅力英尺
lb-in	pounds-inch（es）	磅力英寸
LBTS	land-based test site	陆上试验台
LCAL	low cost aircraft leasing	低成本飞机租赁
LCC	life cycle cost	全寿命周期费用
LCD	liquid crystal display	液晶显示器
LCF	large cargo freighter	大型货机
LCF	low cycle fatigue	低周疲劳
LCH	light combat helicopter	轻型战斗直升机
LCL	lower control limit	控制下限
LCS	Littoral Combat Ships	（美国）濒海战舰
LDM	life discipline manual	寿命规范手册
LDOC	low direct operation cost	低运营成本
LDV	laser doppler velocimeter	激光多普勒测速仪
LE	leading edge	前缘
LE	low energy	低能量
LEAP	low emissions alternative power	（美国国家航空航天局）低污染备选动力计划
LEAP	leading edge aviation propulsion	（CFM 公司）先进航空推进技术项目
LED	light-emitting diode	发光二极管
LES	large eddy simulation	大涡模拟

LES	Eulerian-Lagrangian two phase large-eddy simulation	欧拉-拉格朗日两相大涡模拟
LFD	large freight door	大型货舱门
LFW	linear friction welding	线性摩擦焊
LGRT	laser gradient rig test	激光梯度台架试验
LH	left hand	左手
LHS	left hand side	左手侧
LH₂	liquified hydrogen	液化氢气
LHT	Lufthansa Technik	汉莎技术公司
LHTEC	light helicopter turbine engine company	(罗罗-霍尼韦尔合资)轻型直升机涡轮发动机公司
LIMING	Shenyang Liming Aero-Engines Group Corporation	沈阳黎明航空发动机集团公司
LLP	life limited parts	有限寿命部件
LMC	least material condition	最少留料情况
LMP	life management plan	寿命管理计划
LMP	line maintenance part	航线维护件
LMS	learning management system	学习管理系统
LMS	least mean square	最小均方法
LMT	layered manufacturing technology	分层制造技术
LNG	liquefied natural gas	液化天然气
LOH	light observation helicopter	轻型侦察直升机
LoI	letter of intend	意向书
LOPS	low oil pressure sensor	低油压传感器
LP	logistics provider	后勤保障供应商
LP	low pressure	低压
LPC	low pressure compressor	低压压气机
LPG	liquidized petrolic gas	液化石油气
LPP	lean-premixed-prevaporised combustion	贫油预混气相燃烧
LPSOV	low pressure shut-off valve	低压燃油切断阀
LPT	low pressure turbine	低压涡轮
LRI	line replaceable item	外场可换零件
LRIP	low rate initial production	小批初始生产

LRIS	life and rotor integrity strategy	寿命和转子一体化战略
LRM	line replaceable module	航线可换模块
LRR	lot reject rate	批退率
LRU	line replaceable unit	航线可更换设备
LSA	light sport aircraft	轻型运动飞机
LSL	lower specification limit	规格下限
LSND	low stress no distortion	低应力无变形
LSOP	lubrication and scavenge oil pump	润滑与回油泵
LSP	laser shock processing	激光冲击处理
LSPT	low speed power turbine	低速动力涡轮
LT	low tension	低张力
LTA	long-term agreement	长期合作协议
LTFC	low-temperature fuel cell	低温燃料电池
LTO	landing and take off	着陆和起飞
LTS	Lufthansa Technics Shenzhen	深圳汉莎技术有限公司
LTSA	long-term service agreement	长期售后服务协议
LuFo	Luftfahrt Forschung Program	(德国)航空研究计划
LUH	light utility helicopter	轻型通用直升机
LUVI	laser ultrasonic visualizing inspector	激光超声波可视化检测仪
LVDT	linear variable differential transformer	线性可调差动变压器
LWC	liquid water content	液态水含量

m	metre	米
Ma	Mach number	马赫数
MA	morphing aircraft	(美国)变形体飞机
MAM	manufacturing alteration memorandum	生产变更备忘录
MAPL	modular affordable product line	模块化成本可负担产品线
MAPS	multiple application propfan study	桨扇发动机多重应用研究
MAR	manufacturing alteration request	生产变更请求
MARI	membrane-assistant resin infusion	薄膜辅助液态成型技术(奥地利 FACC 公司)

MAS	manufacturing acceptance sheet	生产许可证
MATV	multi-axis thrust vectoring	多轴推力矢量
MAV	micro air vehicle	微型无人飞行器
MAW	mission adaptive wing	任务自适应机翼
MBD	model based definition	基于模型的设计思想
MBO	management by objectives	目标管理
MBOM	manufacturing bill of materials	制造物料清单
MCAS	maneuvering characteristics augmentation system	操纵特性强化系统
MCD	magnetic chip detector	磁塞铁屑探测器
MCFC	molten carbonate fuel cell	熔融碳酸盐燃料电池
MCL	maximum climb	最大爬升
MCPH	maintenance cost per hour	每小时维修成本
MCPP	manufacturing critical part plan	关键部件生产计划
MCR	manual change request	手册变更要求
MCR	manufacturing change request	生产变更申请
MCR	maximum cruise rating	最大巡航推力状态
MCS	maneuver control system	操纵控制系统
MCT	maximum continuous	最大连续
MD	McDonnell Douglas	(美国)麦克唐纳-道格拉斯公司
MDA	manual draft approval	手册初审
MDA	modification & design approval	重大设计改装批准书
MDAU	maintenance data acquisition unit	维护数据采集装置
MdDA	Ministero della Difensa (Italian)	(意大利)国防部
MDU	manual drive unit	手动装置
ME	manufacturing engineering	制造工程
MEA	more electric aircraft	多电化飞机
MEC	main engine controller	发动机主控制器
Mech	mechanical	机械的
MEDA	maintenance error decision aid	维修差错决断法(波音)
MEE	more electric engine	多电化发动机
MEL	minimum equipment list	最低设备放行
MEM	maintenance and engineering management	维修与工程管理

MEMS	maintenance error management system	维修差错管理系统
MEMS	micro electro mechanical systems	微电动机械系统
MEW	maximum empty weight	最大空重
MFD	multi-function display	多功能显示器
MFI	multifunctional frontline fighter	(俄罗斯)多用途前线战斗机计划
MFVT	mixed flow vectored thrust	混合流矢量推力
MGB	main gear box	主齿轮箱
MHI	Mitsubishi Heavy Industries	(日本)三菱重工业公司
MI	manufacturing instruction	工艺说明
MIL-STD	military standard (USA)	(美国)军队标准
MIM	material increase manufacturing	增材制造
min	minute	分钟
MIS	management information system	管理信息系统
MIS	material inspection specification	材料检验规范
MISC	miscellaneous publication/brochure	各种出版物和小册子
MIT	Massachusetts Institute of Technology	麻省理工学院
MJ	Megajoule	兆焦耳
MLC	manufacturing laboratories catalogue	工艺实验室目录
MLP	manufacturing launch plan	生产启动计划
MLS	materials laboratory services	材料实验室的服务
MLW	maximum landing weight	最大着陆重量
MM	maintenance manual	维护手册
mm	millimetre	毫米
MMA	multimission maritime aircraft	多用途海上飞机
MMC	maximum material condition	最大实体尺寸
MMC	metal matrix composite	金属基复合材料
MMM	materials & mechanical methods	(罗罗公司设计手册)材料和结构计算方法
MMRCA	medium multi-role combat aircraft	中型多用途作战飞机
MMS	mirror milling system	蒙皮镜像铣切系统
Mn	Mach number	马赫数
MNWP	maximum normal working pressure	最大正常工作压力
MO	manufacture order	生产单

MoD	Ministry of Defence（UK）	（英国）国防部
MOD	modification	修改
MoDQAR	ministry of defence quality assurance representative	（英国）国防部质保代表
ModStd	modification standard	修正标准
MOE	maintenance organisation exposition（EASA Part 145）	维修组织说明（欧洲适航局145部）
MOEE	more electric engine	多电化发动机
MoM	method of manufacture	制造方法
MoM	middle of the market	中间市场概念（波音）
MON	mixed oxides of nitrogen	混合氮氧化物
MOP	measures of performance	绩效考评标准
MOP	memorandum of policy	政策大纲
MOPS	main oil pressure sensor	主滑油管压力传感器
MOTS	main oil temperature sensor	主滑油管温度传感器
MoU	memorandum of understanding	谅解备忘录
MPA	maritime patrol aircraft	海上巡逻机
mph	miles per hour	英里每小时
MPI	magnetic particle inspection	磁粉检测
MPI	manufacturing process instruction	初始生产工艺
MPL	master parts list	主要零件表
MPS	material and process specifications	材料和工艺规范
MQI	major quality investigation（Rolls-Royce）	重要质量调查（罗罗）
MQL	minimal quantities of lubricant	最少切削液
MR	mixed reality	混合现实技术
MR	maximum reverse	最大反向
MR&O	maintenance repair & overhaul	维护、维修和大修
MRA	maritime reconnaissance aircraft	海上侦察机
MRA	material return authorisation	退货审批
MRA	material review application	材料应用审核
MRB	material review board	材料审核委员会
MRB	maintenance review board	维修审核委员会
MRCA	multi-role combat aircraft	多用途战斗机
MRI	material return instruction	退货通知

MRJ	Mitsubishi Regional Jet	三菱支线客机(三菱重工,日本)
MRMS	mission ready management solutions	(罗罗公司军用发动机)备勤管理方案
MRO	maintenance repair and overhaul	维护、维修和大修
MRO	material release order	材料放行单
MRP	material requirements/resource planning	材料要求和资源计划
MRPC	manufacturing responsible party company	负责生产制造的伙伴公司
MRT	multi role transport	(德国)多用途运输机
MRTT	multi-role tanker transport	(德国)多用途加油运输机
MS	material specification	材料规范
MSARA	maritime search and rescue aircraft	海上搜索救援飞机
MSDS	materials safety data sheets.	材料安全数据表
MSE	measurement system evaluation	测试系统评估
MSPP	manufacturing sensitive part plan	灵敏部件生产计划
MSR	main supply route	主要供货渠道
MSRR	material specification Rolls-Royce	罗罗公司材料规范
MST	magnet systems technology	磁力系统技术
MT	microturbine	微型涡轮发动机公司
MT	manufacturing technique	制造技术
MTBF	mean time between failures	平均无故障时间
MTBM	mean time between maintenance	平均维修间隔时间
MTBO	mean time between overhaul	平均翻修寿命
MTBUR	mean time between unscheduled removal	平均非计划拆换间隔时间
MTC	marine total care	船舶产品包修协议(罗罗)
MTE	marine trent engine	舰船用遄达发动机
MTF	mid tandem fan	中置风扇
MTF	mid turbine frame	中置涡轮框架
MTI	manufacturing technique instruction	制造技术规定(罗罗)
MTI	moving target indicator	移动目标显示器
MTM	material test methods	材料测试方法
MTO	maximum take-off	最大起飞
MTOT	maximum take-off thrust	最大起飞推力
MTOW	maximum take-off weight	最大起飞重量

MTP	manufacturing technical package	制造技术工作包
MTTF	meantime to failure	平均故障前时间
MTTR	mean time to repair	平均维修时间
MTU	Motoren Und Turbinen Union Munchen GmbH	(德国)(原称慕尼黑发动机涡轮联合公司)摩天宇航空发动机公司
MVA	market value added	市场增加值
MWAA	Metropolitan Washington Airports Authority	美国华盛顿市区机场管理局
MWP	modern working practices	现代工作实施
MZFW	maximum zero fuel weight	最大零油重量

N	Newton	牛顿
N/A	not applicable	不适用
N1	low pressure shaft speed	低压转子转速
N2	high pressure shaft speed (2 shafts), Intermediate shaft speed(3 shafts)	高压转子转速(两轴),中压转子转速(三轴)
N3	high pressure shaft speed (3 shafts)	高压转子转速(三轴)
NAA	National Aeronautic Association	(美国)全国航空协会
NAAR	National Airworthiness Authority Representative	国家适航当局代表
NAB	national accreditation bodies	国家认证机构
NACA	National Advisory Committee for Aeronautics	(美国航空宇航局的前身)国家航空咨询委员会
NADCAP	National Aerospace and Defence Contractors Accreditation Programme	世界航空航天和国防业务承包商认证机构
NAFEC	National Aviation Facilities Experimental Center	国家航空设施实验中心
NAI	nacelle anti-ice	短舱防冰
NAIA	National Aerospace Industry Associations	全国航空航天工业联合会

NAO	National Audit Office of the People's Republic of China	中华人民共和国审计署
NARA	National Archives and Records Administration	国家档案与记录管理局
NAS	National Aerospace Standards	国家航空航天标准
NAS	nuclear acceptance standard	核工业产品验收标准
NASA	National Aeronautics and Space Administration	美国国家航空航天局
NASM	National Air and Space Museum	国家航空航天博物馆
NASP	National Aero-Space Plane	国家航天计划
NATA	National Air Transportation Association	(美国)全国航空运输联盟
NATO	North Atlantic Treaty Organisation	北约组织
NATR	nozzle acoustic test rig	尾喷声学台架试验
NAV	nano air vehicle	超微型无人飞行器
NAVLAP	national voluntary laboratory accreditation program	(美国)国家实验室义务认证计划
NBA	nickel based alloys	镍基合金
NBAA	National Business Aviation Association	(美国)全国公务航空联盟
NBS	National Bureau of Statistics	(中国)国家统计局
NC	numerical control	数字化控制
NCA	non-conformance authority	受权处理异常零件的人或组织
NCB	National Codification Bureau	(北约)国家编码局
NCR	non-conformance report	异常零件报告
NCs	Non-Compliances	不合格
NDE	non-destructive examination	无损检验
NDI	non-destructive inspection	无损检测
NDT	non-destructive testing	无损测试
NDTS	non-dcstructive testing specification	无损测试规范
NEFMO	NATO Eurofighter Management Organisation	北约欧洲战斗机管理机构
NEO	new engine option	换发方案(空客)
NEPA	nuclear energy for the propulsion of aircraft	核能飞机动力

NETMA	NATO Eurofighter and Tornado Management Agency	北约欧洲战斗机和狂风战斗机管理局
NEWAC	NATO Electronic Warfare Advisory Committee	北约电子战咨询委员会
NEWAC	New Aero Engine Core Concepts	(欧洲)新型航空发动机核心机方案
NFAC	national full-scale aerodynamics complex	国家全尺寸空气动力学综合设施
NFCF	notice for changing forecast	更改预估量的通知
NFH	NATO frigate helicopter	北约护卫舰载直升机
NFRP	Singapore Next Fighter Replacement Program	新加坡战斗机更新项目
NGATS	next generation air transportation system	(美国)下一代航空运输系统
NGLT	Next Generation Launch Technology	(美国国家航空航天局)下一代发射技术
NGS	nitrogen generation system	氮气发生器
NGV	nozzle guide vane	导向器叶片
NHA	next higher assembly	次高组合件
NIA	National Institute of Aerospace	国家航空宇航研究所
NiAl	nickel aluminide	铝化镍
NIST	National Institute Standards and Technology	(美国)国家标准与技术研究院
NLF	natural laminar flow	自然层流
Nm	Newton metre	牛顿米
NMA	new midsize aircraft	新型中型客机
NMCS	non mission capable supply	(军机)停机待件
NMS	nuclear manufacturing services	核工业生产制造服务
NO	nitric oxide	一氧化氮
NOS	not otherwise specified	除非另有指定
NOx	nitrogen oxides	氮氧化物
NOx	oxides of nitrogen	氧化氮
NPI	new production introduction	引入新产品
NPNP	no purchase order no payment	不下采购单,不付款
NPRM	notice of proposed rulemaking	建议标准通告

NPRM-Stage 5	notice of proposed rulemaking-stage 5	建议第五阶段标准
NPSS	numerical propulsion system simulation	数字推进系统仿真
NQAA	National Quality Assurance Authority	(英国)国家质量保证局
NQAR	national quality assurance representative(s)	国家质量保证代表
NRC	non-recurring cost	一次性成本
NRC	National Research Council	美国国家研究委员会
NRE	non-recurring engineering	一次性工程
NREC	non-recurring engineering cost	一次性工程成本
NRV	non return valve	单向阀
NSA	new small aircraft	新型支线机项目(波音)
NSMS	non intrusive stress measurement system	非插入式应力测量系统(普惠)
NSN	NATO stock number	北约(13位数)物料编号
NTS	notice to suppliers	给供应商的通知
NTSB	National Transport Safety Bureau	美国国家运输安全署
NTSB	National Transportation Safety Board	国家运输安全委员会
NVFEL	National Vehicle and Fuel Emissions Laboratory	国家飞行器与燃油排放实验室
NVG	night vision goggles	航空夜视镜
NVM	non volatile memory	非易失性内存
NYEX	new york stock exchange	纽约证券交易所

O&M	overhaul & maintenance	大修与维护
OART	office of advanced research and technology	先进研究与技术办公室
OAS	outer air seal	外气封
OASIS	online aerospace supplier information system	在线航空航大供应商信息查询系统
OAST	office of aeronautics and space technology	航空宇航技术办公室
OAT	outside air temperature	外部空气温度

OAV	organic air vehicles	（美国计划研制的垂直起落）无人机
OBU	operational business unit	业务运营单位
OCM	oil control module	滑油控制单元
OD	outer diameter	外径
ODMS	oil debris monitoring system	滑油碎屑监控系统
ODS	oxide dispersion strengthened	氧化物弥散强化
OE	original equipment	原始设备（发动机）
OEB	operations engineering bulletin	运营技术通报
OEDB	Omsk Engine Design Bureau	（俄罗斯）鄂木斯克发动机设计局
OEE	overall equipment effectiveness	设备总效能
OEI	one engine inoperative	单发失效
OEM	original equipment manufacturer	原始设备制造商
OEW	operating empty weight	空载重量
OFDP	oil filter differential pressure sensor	滑油过滤器压差传感器
OGV	outlet guide vanes	出口导向器
OH	occupational health	职业健康
OHSAS	occupational health and safety assessment series（ISO）	职业健康与安全管理体系
OI	operating instructions	操纵手册
OII	other installation item(s)	其他安装相关内容
OLS	oil level sensor	滑油位传感器
OMAT	overhaul materials（Consumable）manual	（可消耗性）大修物料手册
OMC	organic matrix composite	有机复合材料
ONERA	Office National d'Etudes et Recherches Aérospatiales	法国国家航空航天研究院
OOPM	out of phase maintenances	不定期维护
OPEC	Organization of Petroleum Exporting Countries	石油输出国组织
OPR	overall pressure ratio	总增压比
OPRR	operational and production readiness review	运营和生产认证审核
OPU	oil pump unit	滑油泵

OQA	output quality assurance	出货质量保证
OQC	outgoing quality control	出货质量检验
OSU	overspeed and splitter unit	超速和隔板装置
OTAD	oil tank access door	油箱门
OTC	order cycle time	订购循环时间
OTDF	outlet temperature distribution factor	出口温度分布系数
OTWEM	over the wing engine mount	机翼上部装发动机
OVB	outer V blade	外 V 型叶片
OVT	all-axis nozzles vectoring thrust	(俄米格公司的)全向推力矢量
OWC	on-wing care	在翼维护
oz	ounce	盎司

P	power	功率
p	pressure	压力
P&W	Pratt and Whitney	(美国)普惠发动机公司
p. a.	per annum	每年
P/L	payload	有效载荷
P/N	part number	料号,零件号码
P20	total inlet air pressure	进气道空气总压
P26	HP compressor entry pressure	高压压气机入口压力
P2P	peer-to-peer	终端对终端(互联网技术)
P30	HP compressor exit pressure	高压压气机出口压力
P50	LP turbine exit pressure	低压涡轮出口压力
PA	picric acid	苦味酸(水溶性聚合物)
PA	programme agreement	项目合作协议
Pa	Pascal	帕斯卡
PAI	Piaggio Aero Industries	(意大利)比亚乔航空工业公司
PAMB	ambient pressure	环境压力
PanAm	Pan American World Airways	泛美航空公司

PATlimits	production acceptance test limits	产品合格试验极限
Pax	passengers	乘客
Pb	burner pressure	火焰筒压力
PBH	power by hydraulic	功率液传
PBIT	profit before interest and tax	息税前盈利
PBL	performance based logistics	以业绩为准的后勤保障
PBN	performance based navigation	基于飞机性能的导航
PBT	persistent，bio-accumulative and toxic	持久性、生物蓄积性和毒性
PBW	power by wire	功率电传
PCA	Presidency of Civil Aviation	(沙特阿拉伯王国)民航局
PCA	product/project change authority	产品/项目变更审批部
PCA	product control authority	产品控制部门
PCA	propulsion-controlled aircraft	(美国国家航空航天局)用推进系统控制的飞机
PCB	printed circuit board	印刷电路板
PCB	plenum chamber burning	增压室燃烧
PCB	printed circuit board	印刷电路板
PCB	product change board	产品变更委员会
PCD	process/product control document	工艺/产品控制文件
PCDE	project chief design engineer	项目总设计师
PCE	post-certification engineering	取证后工程
PCE	pre-cooler exhaust	预冷器热气排放
PCE	project chief engineer	项目总工程师
PCE	project cost estimate	项目成本预估
PCF	precising casting facility	精铸车间
PCIT	product conformance improvement team	产品一致性改善组
PCL	predicted cyclic life	预估循环寿命
PCN	process change notice	工序改动通知
PCPs	process critical parameters	关键过程参数
PCR	preliminary concept review	初级方案评审
PCR	problem change report	问题变化报告
PCU	propeller control unit	螺旋桨控制器
PDA	phase Doppler analyzer	相位多普勒分析仪
PDD	product definition document	产品定义文件

PDE	pulse detonation engine	脉冲爆震发动机
PDM	product data management	产品数据管理
PDP	primary design review	初级设计审核
PDS	process data sheet	流程数据表
PDSA	plan，do，study，act	计划、执行、检查、调整
PDU	power drive unit	电源驱动装置
PE	process excellence	精益流程
PEM	proton exchange membrane	质子交换膜
PEM	power electronic module	功率电子模块
PEMFC	proton exchange membrane fuel cell	质子交换膜燃料电池
PEST	political，environmental，social and technological influence	政治,环境,社会和技术的影响
PFCS	primary flight control system	飞行控制主系统
PFC	perfluorocarbon	全氟化碳
PFI	private financing initiative	私人融资计划
PFMEA	process failure modes and effects analysis	故障模式和影响分析流程
PFRT	preliminary flight rating test	首次飞行状态试验
PGM	platinum group metals	铂族金属
PHM	prognostics and health monitoring	诊断与健康监测
PHMU	prognostics and health management unit	诊断与健康管理单元
PI	performance improvement	性能改善
PIANO	project interactive analysis and optimization	(初级飞机设计程序)互动式项目分析和优化
PIB	procurement information booklet	采购信息手册
PIC	product integrity committee	产品完整性委员会
PICD	physical interface control document	物理界面控制文件
PILM	product innovation and lifecycle management	产品开发与全寿命周期管理(罗罗)
PIM	Product Introduction Management	产品推介管理
PIMGR	product introduction management gateway review	产品推介管理过门审核
PIP	product introduction process	产品推介流程
PIR	process improvement request	流程改善请求
PIS	power indicating sensor	功率显示传感器

PLA	Chinese People's Liberation Army	中国人民解放军
PLA	power lever angle	功率杆角度
PLASMAERO	useful PLASMA for aerodynamic control	(欧盟)实用等离子体气动控制项目
PLF	pressure loss factor	压力损失系数
PLM	product life-cycle management	产品全生命期管理
PLMBP	product lifecycle management best practice	产品全生命期管理最优方法
PLMOP	product lifecycle management operating procedure	产品全生命期管理运作程序
PLR	partial least-squares regression	偏最小二乘回归分析
PM	preventive maintenance	预防性维护
PMA	parts manufacturer approval	零件生产商的认证
PMA	parts manufacturing approval	零部件制造批准
PMA	permanent magnet alternator (engine dedicated generator)	永磁交流发电机(发动机专用发电机)
PMBLM	permanent magnet brushless motor	永磁无刷电机
PMC	polymer matrix composite	聚合物基复合材料
PMC	power management control	动力管理控制
PMC	production & material control	生产和物料控制
PMP	preventative maintenance plan	预防性维护计划
PMR	preliminary material review	初级物料审核
PMR	polymerization of monomer reactants	单体反应物聚合
PMS	power management system	动力管理系统
PNdB	perceived noise in decibels	感知噪声分贝
PNL	perceived noise level	可感知噪声级
Po	pressure at engine intake	发动机进气口压力
PO	purchase order	订单
PO	purchasing order	采购订单
POA	power optimised aircraft	动力优化飞机
POA	production organisation approval (EASA Part 21)	(欧洲航空安全局第 21 部)生产机构审批
POGO	project on government oversight	(美国)政府项目监督部门
POI	power-off-immersed	关油门状态
POR	power-off-retracted	撤销关油门指令

PowerJet	A joint venture of SAFRAN and NPO	（法国斯奈克玛公司与俄罗斯土星公司的合资）喷气动力公司
PPsheet	problem presentation sheet	问题说明表
PPC	production plan control	生产计划控制
PPD	project policy document	项目政策文件
PPDP	performance and personal development plan	业绩与个人发展计划
PPM	parts per million	百万分率
PPP	purchasing policy & practice	采购方针和方法
PPS	power plant station	发电站
PPS	power plant system	动力装置系统
PQDR	product quality deficiency report	产品质量缺陷报告
PR	pressure ratio	压力比
PRD	product requirements document	产品规范
PRD	project requirements document	项目规范
PRF	pulse repetition frequency	脉冲重复频率
PRG	production readiness guidelines	生产准备指南
PRI	performance review institute	绩效考核机构
ProcPD	procurement policy document	采购策略文件
PRSOV	pressure regulating shut-off valve	压力控制截流阀
PRT	propeller research tunnel	研究螺旋桨的风洞
PRV	pressure regulating valve	压力控制阀
PS	pressure side	压力面
PS	process specification	工艺规范
PSC	parts service centre	零件服务中心
PSC	performance seeking control	寻找并控制最佳性能状态
PSCL	predicted safe cyclic life	预估安全循环寿命
psi	pound per square inch	磅每平方英寸
psia	psi absolute	绝对压力（磅每平方英寸）
psid	psi differential	压差（磅每平方英寸）
psig	psi gauge	传感器测量压力（磅每平方英寸）
PSL	（NASA）propulsion systems laboratory	（美国航空航天局）推进系统实验室

PSQE	plant supplier quality engineer	工厂供应商质量工程师
PSR	product strategy review	产品战略审核
PSSA	propulsion system safety board	推进系统安全理事会
PSSR	project sub-system requirement	项目子系统要求
PSU	power supply unit	动力供应装置
PTA	preferential trade agreement	优惠贸易协定
PTA	preferential trade area	优惠贸易区
PTA	propfan test assessment	桨扇发动机测试评估
PTBM	preliminary type board meeting	型号委员会预备会议
PTC	product technical controller	产品工艺管理员
PTIT	power turbine inlet temperature	动力涡轮进口温度
PTL	project team leader	项目组长
PTS	product test specification	产品检测规范
PVD	physical vapor deposition	电子束物理气相沉积
PVS	portable video system	便携式视频系统
PW	Pratt and Whitney	(美国)普惠发动机公司
PW	pulsed wave	脉冲波
PWC	Pratt & Whitney Canada Corporation	普惠加拿大公司(加普惠,加拿大)
PWR	pressurized water reactor	压水反应堆

Q/R/S	quality/reliability/service	质量/可靠度/服务
Q/STOL	quiet short take-off and landing	静音短距起降
QAM	quadrature amplitude modulation	正交调幅
QAP	quality assurance plan	质量保证计划
QAPI	quality assurance procedure instruction	质保流程条例
QAR	quality assurance representative	质量保证代表
QAS	quality acceptance standard	质量验收标准
QASNDT	quality acceptance standard non-destructive testing	无损检测质量验收标准
QAT	quiet airplane technology	静音飞机技术

QC	quality control	质检
QC	quality-assurance capability	质量保证能力
QCDR	quality，cost，delivery，responsiveness	质量、成本、交货、反应
QCGAT	quiet clean general aviation turbofan	静音低排放通航发动机
QCS	quiet climb system	静音爬升系统
QCSEE	quiet clean short-haul experimental engine	静音低排放短途实验发动机
QCTP	quality control test procedure	试验的质量控制流程
QE	quality escape	质量逃逸
QEC	quick engine change	快速发动机更换件
QECU	quick engine change unit	快速发动机更换装置
QEP	quiet engine program	静音发动机计划
Q-Fan	quiet-fan	静音风扇
QHSF	quiet high speed fan	静音高速风扇
QMS	quality management system	质量管理系统
QN	quality notification	质量通报
QOP	quality operating procedure	质量管理程序
QP	quality plan	质量计划
QR	quick release	快速发布
QRC	quality resolution co-ordinator	质量解决协调人
QRL	quality renaissance leader	质量复兴项目主管
QS	quality standard	质量标准
QSRA	quiet short-haul research aircraft	静音短距研究机
QSST	quiet supersonic transport	（美国 Aerion 集团）静音超音速客机
QT	qualification test	定型试验
QTD	quiet technology demonstrator	静音技术验证机（波音）
QTD2	quiet technology demonstrator 2	第二号静音技术验证机（波音）
Qty	quantity	数量
QUESTOL	quiet experimental short takeoff and landing	静音实验性短距起降

R&D	research and development	研究与开发
R&O	repair and overhaul	修理和大修
R&T	research and technology	研究和技术
RAAF	Royal Australian Air Force	澳大利亚皇家空军
Rad	radial	径向
RALS	remote augmented lift system	远距增升系统
RAMSES	reduced acoustic mode scattering engine system duct	低噪声散射发动机管道系统
RASM	revenue per available seat mile	每座英里收入
RAT	ram air turbine	冲压空气涡轮
RATO	rocket assisted take-off	火箭辅助起飞
RATTLRS	revolutionary approach to time critical long range strike	（美国）远程打击时间敏感目标的导弹项目
RB	rolls by-pass	罗尔斯旁路
RBC	radiation barrier coating	防热辐射涂层
RBC	rear bearing chamber	后部轴承腔
RBCC	rocket based combined cycle	火箭基组合循环
RBSN	reaction-bonded silicon nitride	反应烧结氮化硅
RBT	resistance bulb thermometer	电阻温度传感器
RC	recurring cost	经常性成本
RCA	rolling circle arc	滚动圆弧
RCA	root cause analysis	根本原因分析
RCC	reinforced carbon-carbon	增强碳纤维复合材料
RCM	reactor core materials	反应堆芯材料
RCP	resolve customer problem	解决客户问题
RCP	root cause report	根本原因报告
RDC	remote data concentrator	远程数据采集器
RDM	research, development and manufacturing	研发和生产
RDS	radial driving shaft	径向传动轴
REACH	registration, evaluation and authorisation of chemicals	（欧盟）化学品的注册评估和审批草案

rel	related	相关的
RELT	Rolls E. L. Turbofans Ltd.	（罗罗子公司负责发动机支持和服务）罗尔斯涡扇发动机公司
RESA	request for engineering source approval	要求批准工程资源
RevCon	revolutionary concepts in aeronautics RG	航空记录组的革命性方案
RF	range factor	航程因子
RF	reserve factor	安全系数
RFA	request for alteration	变更申请
RFC	request for comment	请求注释
RFI	request for information	信息征询书
RFI	resin film infusion	树脂膜熔渗成形
RFID	radio frequency identification	射频识别技术
RFID tags	radio frequency identification tags	射频识别标签
RFO	request for offer	要求报价
RFP	request for proposal	招标
RFQ	request for qualifications	质量要求
RFQ	request for quotation	询价
RFT	right first time	一次成功
RFV	request for variation	变更申请
RH	right hand	右手
RHS	right hand side	右手侧
RHFCU	re-heat fuel control unit	加力燃油控制器
RIA	runway independent aircraft	超短距起降飞机
RII	required inspection items	要求检验的项目
RIP	reach implementation project	（欧盟化学品注册评估和审批政策）实施项目
RIT	rotor inlet temperature	转子进口温度
RITA	Research and Innovative Technology Administration	（美国交通运输部）研究和技术创新局
RJ	regional jet	支线飞机
RMA	returned material approval	退货验收
RN	release note（CoC）	产品发放注释单

ROC	return on capital	资本利润率
ROCE	return on capital employed	已动用资本回报率
ROM	rough order of magnitude	粗略估计
ROS	repair of spares	备件修理
ROV	remotely operated vehicle	遥控车
ROW	rest of the world	世界其他地方
RPFH	rate per flight hour	按小时付费
RPK	revenue passenger kilometer	每旅客人公里收费
rpm	revolutions per minute	每分钟转速
RPN	risk priority number	风险顺序数
RPS	rationalised process specification	优化的流程规范
RPV	remotely piloted vehicle	遥控飞行器
RQL	rich-quench-lean combustion	富油-猝熄-贫油燃烧
RQS	rationalised quality standards	罗罗公司质量标准
RQSC	rationalised quality standard-component specific	罗罗公司特殊部件质量标准
RQSG	rationalised quality standard-generic	罗罗公司通用质量标准
RQSP	rationalised quality standard-process	罗罗公司工艺质量标准
RQST	rationalised quality standard-component type	罗罗公司部件型号质量标准
RR	Rolls-Royce plc.	（罗尔斯·罗伊斯股份公司）罗罗公司
RRA	regional regulatory authority（Rolls-Royce）	区域性管理机构（罗罗）
RRB	Rolls-Royce Brazil	罗罗巴西公司
RRC	Rolls-Royce Corporation	罗罗美国公司
RRC（R&O)	Rolls-Royce Canada Repair & Overhaul	罗罗加拿大修理和大修公司
RRD	Rolls-Royce Deutschland Ltd & Co. KG	罗罗德国有限公司
RRECSB	Rolls-Royce energy customer service business	罗罗能源客户服务业务
RREGT	Rolls-Royce Energy Gas Turbines Canada	罗罗能源加拿大燃机公司
RREP&PS	Rolls-Royce Energy Package & Power Systems	罗罗能源系统集成和动力系统公司

RRES	Rolls-Royce Energy Systems	罗罗能源系统公司
RRES	Rolls-Royce engineering specification	罗罗技术规范
RRESO	Rolls-Royce Engine Services Oakland	罗罗(美国)奥克兰能源服务公司
RRJ	Russia Regional Jet	俄国支线喷气机
RRM	Rolls-Royce marine	罗罗公司舰船业务
RRMES	Rolls-Royce marine electrical systems	罗罗公司舰船电气系统
RRMP	Rolls-Royce marine power	罗罗公司舰船动力
RRMPOL	Rolls-Royce Marine Power Operations Ltd.	罗罗公司舰船业务公司
RRMS	Rolls-Royce marine systems	罗罗公司舰船系统
RRMSP	Rolls-Royce managed service provider	罗罗公司管理的服务商
RRN	Rolls Royce North America Inc	罗尔斯·罗伊斯美国公司
RRPE	Rolls-Royce Power Engineering Ltd.	罗罗公司动力工程公司
RRPS	Rolls-Royce production system	罗罗公司生产系统
RRQMS	Rolls-Royce quality management system	罗罗公司质量管理系统
RRS	risk revenue sharing	风险收益共担
RRSP	risk and revenue sharing partner	风险收益共担伙伴
RRSP	risk and revenue sharing partnership	风险收益共担伙伴关系
RRTM	Rolls-Royce Turbomeca	罗罗透博梅卡合资公司
RRUK	Rolls-Royce (UK Aerospace)	英国罗尔斯·罗伊斯公司
R/S	rotating stationary blade	动/静叶
RSD	reliability and safety documentation	可靠性和安全性文件
RSP	risk sharing partner	风险共享伙伴
RSPL	recommended spare parts list	首批推荐备件清单
RTA	revolutionary turbine accelerator	革命性涡轮加速器
RTC	reduced temperature configuration	降温构型叶片
RTCU	range temperature control unit	温度范围控制器
RTD	resistance temperature device	电阻测温器
RTDF	radial temperature distribution factor	径向温度分布数
RTM	resin transfer molding	树脂传递成型
RTP	reduce to produce	降耗减重生产
RTPE	route to process excellence	卓越流程项目指南(罗罗)
RTS	runtime system	(软件)运行系统

RTT	return-to-tank valve	回油箱阀
RVDT	rotary variable differential transformer	旋转可调差分变压器
RVP	reid vapour pressure	雷德蒸发压力
RVR	runway visual range	跑道能见度
RW	ramp weight	最大滑行重量
RWD	rearward	向后

s	second	秒
S/N	serial number	序列号
S&OP	sales & operations planning	销售和运营计划
S/E	service engineer	服务工程师
S/S	sample size	样本大小
S1000D	International Specification for the Production of Technical Publications	(欧美联合)产品技术出版物国际规范
SA	surface area	表面积
SABRe	supplier advanced business relationship	(罗罗公司与)供应商的先进业务关系
SAC	standard annular combustor	标准环型燃烧室
SACOC	surface air cooled oil cooler	表面气冷滑油冷却器
SADRI	Shenyang Aeroengine Design and Research Institute	沈阳航空发动机研究院
SAE	society of automotive engineers	汽车工程师协会
SAESL	Singapore Aero Engine Services Limited.	新加坡航空发动机服务有限公司
SAF	sustainable aviation fuel	(环保或绿色)可持续航空燃料
SAGB	step-aside gearbox	测装齿轮箱
SAM	surface-to-air missile	地空导弹
SAM	sound absorption material	吸声材料
SAN	System Anormaly Note	系统异常记录
SAP	system application product	系统应用和产品

SAR	search and rescue	搜寻和营救
SAR	specific air range	比航程
SAR	synthetic aperture radar	合成孔径雷达
SAS	secondary air system	辅助空气系统
SAV	starter air valve	起动机气阀
SB	service bulletin	服务通报（罗罗）
SBAC	Society of British Aerospace Companies	英国航空企业协会
SBAL	static balance	静平衡
SBIR	small business innovation research	小型商务创新研究
SBJ	speedy business jet	快速公务机
SBP	supplier briefing pack（SABRe）	供应商须知
SBPD	structural by-pass duct	承载结构外涵道
SC	single crystal	单晶
SCAC	Sukhoi Civil Aircraft Corporation	苏霍依民机公司
SCAR	supersonic cruise aircraft research	超声速巡航飞机研究
SCC	supply chain capability	供应链能力
SCCB	software configuration control board	软件配置管理委员会
SCD	source controlled drawing/definition	固定资源图/定义
SCEM	supply chain event management	供应链业务管理
SCM	standard cubic metre	标准立方米
SCM	suppliers chain management	供应链管理
SCMP	system configuration management plan	系统方案管理计划
SCMR	surface combatant maritime rotorcraft	（英国）水面作战海上旋翼机
SCPC	supply chain planning capability	供应链计划能力
SCRAMJET	supersonic combustion ramjet	超燃冲压喷气发动机
SCRIA	supply chain relationships in aerospace	航空航天工业中的供应链关系
SCS	supply chain strategy	供应链战略
SCU	supply chain unit	供应链管理部
SD&Q	supplier development & quality	供应商的开发与质量
SDD	system design and demonstration	系统设计和验证
SDD	system design and development	系统设计和开发
SDD	system development and demonstration	（F135发动机）系统发展和验证

SDIL	supplier development improvement leader	供应商发展改进负责人
SDL	supplier development leader	供应商发展负责人
SDLF	shaft driven lift fan	轴驱动升力风扇
SDP	system design process	系统设计流程
SDRO	standard data records office	标准数据记录办公室
SDS	supplier delivery specification	供应商交货规范
SDSF	short duck separate flow	短涵道分离流
SDT	self-directed team	自主团队
SDURD	system design and unit requirement definition	系统设计和零件要求定义
SEAL	small engine assembly line	小型发动机组装线
sec	second	秒
SEC	software enabled control	软件启动控制
SECTF	small engine component test facility	小发部件试验台
SEM	scanning electron microscopy	扫描电子显微镜
SEMA	Special Electronic Mission Aircraft	特殊电子任务飞机
SFAR	Special Federal Aviation Regulation	美国特种联邦航空条例
SFC	specific fuel consumption	耗油率
SFDM	shop floor data management	车间数据管理
SFE	supplier furnished equipment	供应商安装的设备
SFN	separate-flow nozzle	分离流尾喷管
SFO	serious fraud office	(英国)严重诈骗事件调查办公室
SFW	subsonic fixed wing	亚声速固定翼
SG	specific gravity	比重
SGML	standard generalised markup language	标准通用标示语言
SHFE	small heavy fuel engine	(霍尼维尔公司)小型重油发动机
SHM	structural health monitoring	结构健康检测
shp	shaft horsepower	轴马力
SID	standard instrument departure	标准仪表离场
SIP	standard inspection procedure	检验标准程序
SIS	scheme issue statement	出图说明
SIS	supplier information system	供应商信息系统

SJAC	Society of Japanese Aerospace Companies	日本航空企业协会
SL	see level	海平面
SLISA	sea level international standard atmosphere	海平面国际标准大气
SLA	stereotithography apparatus	光敏树脂快速成型机
SLA	service level agreement	服务水平协议
SLC	shelf life code	储存期产品号
SLEP	service life extension program	（美国空军）发动机使用寿命延长计划
SLIS	strength life and integrity statement	强度寿命及完整性说明
SLPC	single-lever power control	单杆动力控制
SLS	strength and life statement	强度和寿命的定义
SLTOSTD	see level take-off standard	海平面起飞标准条件
SMA	shape memory alloy	变形记忆合金
SMAW	shielded metal arc welding	手工电弧焊
SMC	source method control	源方法控制
SMD	sauter mean diameter	索太尔平均直径
SME	small/medium enterprises	中/小型企业
SMP	service management plan	服务管理计划
SMS	safety management system	安全管理系统
SMT	supplier management team	供应商管理小组
SND	low stress non-deformation	低应力无变形
Snecma	Societe Nationale d'Etude et de Construction de Moteurs d'Aviation	斯奈克玛公司（赛峰集团，法国）
SOC	shut-off cock	停车开关
SOFC	solid oxide fuel cell	固态氧化物燃料电池
SOMS	standard operations management system	标准运营管理系统
SOP	standing operating procedure	标准操纵规定
SORB	Sales and Operations Review Board	销售和运营审查委员会（罗罗）
SOT	stator outlet temperature	涡轮前导向器出口温度
SOV	shut-off valve	切断阀
SOW	statement of work	任务说明书
SO_x	oxides of sulphur	硫氧化物
SP	service provider	服务商

SPC	statistical process control	统计过程控制
SPF	super plastic forming	超塑成型
SPF	Super-Plastically Formed	用超塑成型法制造
SPG	special process group	特定流程组
SPIS	services packaging instruction sheet	包装服务规定表
SPL	sound pressure level	声压级
SPO	system program office	系统计划办公室
SPR	seal pressure ratio	封严压比
SPS	secondary power system	用于军机地面服务的辅助动力系统(由辅助动力装置和两个齿轮箱组成)
SPS	solar power station	太阳能发电站
SPT	stationary plasma thruster	静态等离子推进器
SPTE	special purpose test equipment	专用试验设备
SQA	strategy quality assurance	策略质量保证
SQE	supply quality engineer	供货质量工程师
SQEP	suitably qualified and experienced personnel	有资质和经验的员工
SQM	supplier quality memorandum	供应商质量备忘录
SQT	supply quality team	供货质量组
SR	single-rotation（propfan）	单转子(桨扇发动机)
SRA	strategic research agenda	(欧盟)战略研究计划
SRB	safety review board	安全审查委员会
SRM	scheme review/release meeting	审图/出图会议
SRM	switched reluctance motor	开关磁阻电机
SRS	submarine rescue system	水下救生系统
SRV	submarine rescue vehicle	水下救生艇
SS	specification search	(搜寻特定规格的产品)特寻
SS	suction side	(叶片)吸力面
SSA	special security agreement	专门保密协议
SSA	sub-system accountability	子系统问责制
SSAMC	Sichuan Snecma Aeroengine Maitainance Corporation	四川斯奈克玛航空发动机维修公司
SSBJ	supersonic business jet	超音速公务机

SSDD	sub-system definition document	子系统定义文件
SSG	strategic sourcing group	战略采购组
SSPC	solid state power controller	固态功率控制器
SSRD	sub-system requirements document	子系统要求文件
SSSS	strategic sourcing supplier selection	战略采购供应商的选择
SST	supersonic transport	超声速运输机
SSTJ	single spool turbojet	单轴涡轮喷气发动机
ST	specific tools	专用工具
STAECO	Taikoo（Shandong）Aircraft Engineering Co.，Ltd.	山东太古飞机工程有限公司
STARCO	Shanghai Technologies Aerospace Company Limited	上海科技宇航有限公司
STEM	science，technology，engineering and maths	科学、技术、工程、数学
STEP	supplier total evaluation process	供应商全面评估流程
STG	steam turbine generator	蒸气涡轮发电机
STM	scanning tunneling microscope	扫描隧道显微镜
STOBAR	short take-off but arrested recovery	短距起飞,拦阻降落
STOL	short take-off and landing	短距起降
STOVL	short take-off and vertical landing	短距起飞垂直着陆
STOVL engine	short take-off and vertical landing engine	短距起飞垂直着陆发动机
STP	solar thermal propulsion	太阳能火箭发动机
STSM	sub-tier supplier management	次级供应商管理
STTE	special tools and test Equipment	特殊工具和试验设备
SUAVE	small unmanned aviation vehicle engine	小型无人机发动机
SV	shop visit	返修
SVA	stator vane actuator	导向器作动器
SVHC	substances of very high concern	(欧盟)高度担忧物质
SVO	starter valve open	启动器阀门打开
SVR	shop visit rate	返修率
SVV	system verification vehicle	系统验证机
SWOT	strengths，weaknesses，opportunities，threats	强项,弱项,机会,威胁

| SWPC | Siemens-Westinghouse Power Corporation | 西门子—西屋动力公司 |
| SWPM | solid wood packing material | 实木包装材料 |

T/Rev	thrust reverser	反推装置
T/X	time expiry	期满
T20	total inlet air temperature	进气道空气总温
T26	HP compressor entry temperature	高压压气机进气温度
T30	HP compressor delivery temperature	高压压气机出气温度
TA	triggered assessment	触发性评估
TACC	turbine active clearance control	涡轮间隙主动控制
TAESL	Texas Aero Engine Services Limited	得克萨斯航空发动机服务有限公司
TAI	thermal anti-icing	热除冰
TALON	Technology for Advanced Low NOx	"泰龙"燃烧室（普惠）
Tamb	air ambient temperature	外部空气温度
TAML	Tata Advanced Material Limited	（印度）塔塔先进材料有限公司
TAPS	twin-annular premixing-swirler	双环腔预混涡流（燃烧室）（GE）
TAS	true air speed	真空速
TAT	total air temperature	空气总温
TAT	turn around time	周转时间
T_B	block time	轮档时间
TBA	to be advised	待定
TBA	test bed analysis	试车台分析
TBC	thermal barrier coating	热障涂层
TBCC	turbine based combined cycle	涡轮基联合循环（发动机技术）
TBD	to be determined	待定
TBE	turbine bypass engine	涡轮涵道发动机
TBO	time between overhauls	翻修寿命

TBH	turbine outlet passage	涡轮出口流道
TC	thermocouple	热电偶
TC	time-control capability	进度控制能力
TC	TotalCare	(罗罗公司发动机)包修服务
TC	type certification	型号合格证
TCA	turbine cooling air	涡轮冷却空气
TCA	tast control architecture	任务监控方案
TCA	TotalCare Agreement	(罗罗公司发动机)包修协议
TCA	Transport Canada	加拿大运输部
TCA	technology concept airplane	技术方案验证机
TCAS	traffic collision avoidance system	空中交通防撞系统
TCCA	Transport Canada Civil Aviation	加拿大运输部民航局
TCH	type certificate holder	型号合格证持有者
TCM	thrust control malfunction	推力控制故障
TCP	TotalCare Package	包修协议包
TCR	technical capability review	技术能力评审
TCS	total component support	(汉莎技术公司的)全面部件支持
TCT	type certification team	型号适航证工作组
TDC	top dead centre	顶部驻点
TDI	technical data integration	技术参数一体化
TDTM	technical data tracking monitor	技术数据跟踪监控
TE	trailing edge	(叶片)后缘
TEC	turbine exhaust case	涡轮排气机匣
TechPub	technical publication	技术出版物
TEI	text element identifiers	文本单元标识符
TEMP	temperature	温度
TEN	thrust、efficiency、new technology	推力、效率、新技术(罗罗)
TEPIC	high-temperature，high-strength structural support foam	(美国开发的)高温高强结构支持泡沫材料
TES	total engine support	(汉莎技术公司的)全面发动机支持
TET	turbine entry temperature	涡轮入口温度

TF	task force	特别工作组
TF	turbofan	涡扇发动机
TFN	total fixed nitrogen	全部固态氮
TFP	tailored fibre placement	定制的纤维铺放
TFP	turbo fuel pump	涡轮燃油泵
TFX	tactical fighter experimental	战术战斗机实验机
TGA	thermal gravity analysis	热重分析法
TGL	temporary guidance leaflet	临时指导单
TGS	true ground speed	真地速
TGT	turbine gas temperature	涡轮燃气温度
THOR	testing high-tech objectives in reality	现实中的高科技目标实验（空客 3D 打印飞机）
TI	technical instruction	技术手册
TiAl	titanium aluminide	铝化钛
TIC	turbine intermediate case	涡轮中介机匣
TiGr	titanium-graphite	石墨纤维钛合金层板
TIHE	triggered isomer heating exchanger	（美国）触发同分异构体热交换器
TiMC	titanium matrix composite	钛基复合材料
TIMS	tag information management system	标识信息管理系统
TIMS	technical information management system	技术信息管理系统
TIP	type investigation programme	型号事故调查程序
TIT	turbine inlet temperature	涡轮进口温度
TL	Textron Lycoming	（美国）莱康明公司
TLA	thrust lever angle	推力杆角度
TLD	time limited dispatch	限时遣派
TLM	time limits manual	时效手册
TLP	transient liquid phase	瞬间过渡液相扩散焊
TLR	technical life review	工程寿命评审
TLV	track lock valve	轨道保险阀
TM	Turbomeca	（法国赛峰集团）透博梅卡发动机公司
TM	torque motor	扭矩马达
TMC	titanium matrix composite	钛基复合材料

TMF	thermal fatigue	热疲劳
TN	technical note	技术说明
TOA	technical oversight assessment	技术监督评估
TOAPC	technical oversight assessment-planning & control	计划与控制的技术监督评估
TOAQ	technical oversight assessment-quality	质量的技术监督评估
TOB	terms of business	商务条件
TOC	throttles only control	仅靠推力杆控制
TOFL	take-off field length	起飞跑道长
TOM	total office management	办公室全面管理
TOR	take-off rating	起飞推力设置
TOS	turbine overspeed	涡轮超速
TOS	total operational support	全面运营支持
TOSS	take-off safety speed	安全起飞速度
TOW	take off weight	起飞重量
TP	turbo-prop	涡轮-螺旋桨
TPE	turbo-prop engine	涡桨发动机
TPI	test process instruction	测试工艺
TPM	total productive maintenance	全面生产性维护
TPT	torque pressure transmitter	扭转压力转换器
TQC	total quality control	全面质量控制
TQM	total quality management	全面品质管理
TR	technical report	技术报告
TRA	throttle resolver angle	推力杆解算器角度
TRAS	thrust reverser actuation system	反推力作动系统
TRCU	thrust reverser control unit	反推力控制器
TRD	test requirements document	试验要求文件
TRIZ	theory of inventive problem solution	发明解决问题的理论
TRL	technical readiness level	技术成熟度
TRT	turn round time	周转期
TRU	thrust reverser unit	反推力装置
TS	technical specification	技术规范
TS&O	technical support & operations	技术支持和经营
TSA	transportation security agency	(美国)运输安全局

TsAGI	Central AeroHydrodynamic Institute	(俄罗斯)茹科夫斯基中央空气流体力学研究院
TSD	technical service directory	技术服务手册
TSFC	specific fuel consumption	燃油消耗率
TSM	technical surveillance metallurgist	冶金技术监控专家
TSTF	twin spool mixed turbofan	双轴混合流涡扇发动机
TTH	tactical transport helicopter	战术运输直升机
TTS	total technical support	(汉莎技术公司的)全面技术支持
TU	Tupolev Design Bureau	(俄罗斯)图波烈夫设计局
TUAV	tactical unmanned air vehicle	战术无人机
Turb	turbulence	紊流
TV	technical variance	技术变更
TVC	trapped vortex combustor	驻涡燃烧室
TW	turbine work per mass of air flow	单位流量涡轮功
TWI	The Welding institute	(英国)焊接研究所
TWIC	transportation worker identification credential	(美国)运输工人身份卡
TWP	turbo water pump	涡轮水泵
TX206	materiel acceptance and release form	物料收发单(罗罗)

U/S	unserviceable	无用的
UAC	United Aircraft Corporation	(俄罗斯)联合飞机制造集团
UAL	United Airlines	美联航
UART	universal asynchronous receiver/transmitter	通用同步接收器/发射器
UAV	unmanned aerial vehicle	无人机
UCAR	unmanned combat armed rotor craft	(美国)无人战斗武装旋翼机
UCAV	unmanned combat air vehicle	无人战斗机
UCL	upper control limit	控制上限
UDC	voltage direct current	直流电

UDF	unducted fan	无涵道风扇
UDFE	unducted fan engine	无涵道风扇发动机
UDLM	unscheduled depot level maintenance	非计划内场级维修
UEC	United Engine Corporation	(俄罗斯)联合发动机制造集团
UEET	ultra-efficient engine technology	(美国国家航空航天局的)极高效发动机技术(计划)
UEVM	universal engine vibration monitor	通用发动机振动监测仪
UFFA	universe fan facility adaptation	通用风扇设备适应性
UFIS	universal flight information system	全球飞行信息系统
UHB	ultra high-bypass	超高涵道比
UHBR	ultra high bypass ratio	超高涵道比
UHC	unburned hydrocarbons	未燃碳氢化合物
UHC	unburnt hydrocarbons	未燃烧碳氢化合物
UHF	ultra high frequency	超高频率
UHPE	ultra high propulsive efficiency	(齿轮传动)超高推进效率
UKGAAP	UK Generally Accepted Accounting Principles	英国公认会计原则
UKAS	United Kingdom Accreditation Service	英国测量认证机构
UPS	uninterruptible power supply	不间断电源
USAF	US Air Force	美国空军装备总部
USET	upper stage engine technology	(美国)高级发动机技术项目
USG	United States Government	美国政府
USL	upper specification limit	规格上限
USMC	U.S. Marine Corps	美国海军陆战队
USP	ultrasonic shot peening	超声喷丸
UTAS	United Technologies Aerospace Systems	联合技术公司航空航天系统
UTC	United Technologies Corporation	(美国)联合技术公司
UTRC	United Technologies Research Center	联合技术公司研究中心
UTS	ultimate tensile strength	极限拉伸强度
UWBS	universal wireless backbone system	通用无线主干系统(空客)

V	volts	伏
V/STOL	vertical and short takeoff and landing	垂直与短距起降
VA	vertical audit	垂直评审
VAATE	versatile affordable advanced turbine engine	(美国)负担得起的多用途先进涡轮发动机
VABI	variable area bypass injector	可调面积涵道引射器
VAR	vaccum arc remelting	真空自耗电弧熔烁
VARTM	vacuum assisted resin transfer molding	真空辅助树脂模塑工艺
VBV	variable bleed valve	可调放气阀
V-CAP	vehicle charging and potential	飞行器充电和电压
VCE	variable circle engine	变循环发动机
VDC	voltage direct current	直流电
VDT	variable-density tunnel	变密度风洞
VE	virtual enterprise	虚拟企业
VE	value engineering	价值工程
VEAS	equivalent air speed	当量空速
VERP	virtual enterprise resource planning	虚拟企业资源计划
VFAN	variable fan nazzle	可调风扇导向器
VFDR	variable flow ducted rocket	变流量涵道火箭
VFG	variable frequency generator	变频发电机
VGV	variable guide vane	可调导向器
VHB	very high bypass	极高涵道比
VIAM	All Russia Institute of Aviation Materials	俄罗斯航空材料研究院
VIF	vectoring in flight	飞行矢量
VIGV	variable inlet guide vane	可调进气导向器
Virginia Tech/ VT	Virginia Polytechnic Institute	弗吉尼亚理工大学
VITAL	environmentally friendly aero engine	环保型航空发动机(欧洲联盟资助项目)
VIVACE	value improvement through a virtual aeronautical collaborative enterprise	(欧盟项目)利用虚拟仿真方法实现航空企业增值

VLJ	very light jet	超轻型喷气机
VLTA	very large transport airplane	超大型运输飞机
VM	virtual manufacturing	虚拟制造
V_{MD}	minimum drag speed	最小飞行速度
VMO	variable metering orifice	可调节流油孔
Vorbix	vortex burning and mixing	涡流燃烧与混合
VORV	variable oil reduction valve	可变减油阀
VPD	virtual product development	虚拟产品开发
VPM	virtual product management	虚拟产品开发管理
VPN	virtual private network	虚拟专用网络
VPVB	very persistent and very Bio-accumulative	高持续性和高生物积累性
VR	virtual reality	虚拟现实
VSCE	variable stream control engine	变流路控制发动机
VSCF	variable speed constant frequency	等频变速
VSD	vertical situation display	(飞机)高度状况显示
VSR	validation strategy report	验证策略报告
VSV	variable stator vane	可调静子叶片
VTAS	calibrated air speed	校准空速
VTOL	vertical take-off and landing	垂直起降

W	Watt	瓦
W/L	working line	工作线
WAI	wing anti-ice	机翼防除冰系统
WB	wide body	宽体(飞机)
WBS	work breakdown structure	任务分解结构
WBT	web-based training	网上培训
WEC	Westinghouse Electric Company	(美国)西屋电气公司
WED	whole engine definer	整体发动机方案制定者
WEM	whole engine model	整机模型
WF	weight of fuel	燃油重量

WI	Williams International	（美国）威廉姆斯国际发动机公司
WIP	work in progress	进展中的工作
WLC	whole life cost	全寿命成本
WMCU	water methanal control unit	（高压比发动机）水与甲醛比例控制装置
WOW	weight on wheels	起落架负重
WP	work package	工作包
WPO	work package owner	工作包负责人
WPS	works process specifications	工作流程规范（罗罗）
WSU	wheel spin up	轮子打滑
WTT	customer working together team	与客户协同工作组

XAE	Xian Aero-Engines Corporation	西航公司
XCK	cross check	互相检验
XRA	Xi'an Rolls-Royce Aero Components	西安西罗航空部件有限公司
XRF	X-Ray fluorescence	X 射线荧光分析
XWB	extra wide body （A350）	超宽体飞机（A350）

Yak	Yakovlev Design Bureau	（俄罗斯）雅克飞机设计局
yd	yard	码
YTD	year to date	（指从会计年度开始到报表日的期间）本年度迄今

附录 英制与公制单位
转换中英对照表

量的名称	换算系数(英)	换算系数(中)
Length 长度	$1\,\text{in}=25.4\,\text{mm}$ $1\,\text{ft}=0.3048\,\text{m}$ $1\,\text{mile}=1.609\,34\,\text{km}$ $1\,\text{nautical mile}=1.852\,\text{km}$	1英寸=25.4毫米 1英尺=0.3048米 1英里=1.609 34公里 1海里=1.852公里
Area 面积	$1\,\text{in}^2=645.16\,\text{mm}^2$ $1\,\text{ft}^2=92\,903.04\,\text{mm}^2$	1平方英寸=645.16平方毫米 1平方英尺=92 903.04平方毫米
Volume 体积/容积	$1\,\text{UK fluid ounce}=28\,413.1\,\text{mm}^3$ $1\,\text{US fluid ounce}=29\,573.5\,\text{mm}^3$ $1\,\text{Imperial pint}=568\,261.0\,\text{mm}^3$ $1\,\text{US liquid pint}=473\,176.0\,\text{mm}^3$ $1\,\text{UK gal}=4\,546\,090.0\,\text{mm}^3$ $1\,\text{US gal}=3\,785\,410.0\,\text{mm}^3$ $1\,\text{in}^3=16\,387.1\,\text{mm}^3$ $1\,\text{ft}^3=0.028\,316\,8\,\text{m}^3$	1英制液盎司=28 413.1立方毫米 1美制液盎司=29 573.5立方毫米 1帝国品脱=568 261.0立方毫米 1美制液品脱=473 176.0立方毫米 1英制加仑=4 546 090.0立方毫米 1美制加仑=3 785 410.0立方毫米 1立方英寸=16 387.1立方毫米 1立方英尺=0.028 316 8立方米
Mass 质量	$1\,\text{oz}=28.349\,523\,\text{g}$ $1\,\text{lb}=0.453\,592\,37\,\text{kg}$ $1\,\text{ton (UK)}=1.016\,05\,\text{tonne}$ $1\,\text{short ton}=2\,000\,\text{lb}=0.907\,184\,74\,\text{tonne}$	1盎司=28.349 523克 1磅=0.453 592 37公斤 1英制吨=1.016 05吨 1短吨=2 000磅=0.907 184 74吨
Density 密度	$1\,\text{lb/in}^3=27\,679.9\,\text{kg/m}^3$ $1\,\text{lb/ft}^3=16.0185\,\text{kg/m}^3$	1磅/立方英寸=27 679.9公斤/立方米 1磅/立方英尺=16.018 5公斤/立方米

（续表）

量的名称	换算系数（英）	换算系数（中）
Velocity 速度	$1\,\text{in/min}=0.423\,33\,\text{mm/s}$ $1\,\text{ft/min}=0.005\,08\,\text{m/s}$ $1\,\text{ft/s}=0.304\,8\,\text{m/s}$ $1\,\text{mi/h}=1.609\,34\,\text{km/h}$ $1\,\text{knot}=1.852\,\text{km/h}$	1英寸/分钟=0.423 33 毫米/秒 1英尺/分钟=0.005 08 米/秒 1英尺/秒=0.304 8 米/秒 1英里/小时=1.609 34 公里/小时 1节=1.852 公里/小时=0.514 444 m/s
Acceleration 加速度	$1\,\text{ft/s}^2=0.304\,8\,\text{m/s}^2$	1英尺/秒平方=0.304 8 米/秒平方
Mass Flow Rate 质量流速	$1\,\text{lb/h}=1.259\,98\times10^{-4}\,\text{kg/s}$	1磅/小时=1.259 98×10^{-4} 公斤/秒
Force 力	$1\,\text{lbf}=4.448\,22\,\text{N}$ $1\,\text{kgf}=9.806\,65\,\text{N}$ $1\,\text{tonf}=9\,964.02\,\text{N}$	1磅力=4.448 22 牛顿 1公斤力=9.806 65 牛顿 1吨力=9 964.02 牛顿
Pressure 压力	$1\,\text{in Hg}\,(0.033\,863\,9\,\text{bar})=3\,386.39\,\text{Pa}$ $1\,\text{lbf/in}^2\,(0.068\,947\,6\,\text{bar})=6\,894.76\,\text{Pa}$ $1\,\text{bar}=100.0\,\text{kPa}$ $1\,\text{standard atmosphere}=101.325\,\text{kPa}$	1英尺汞柱=3 386.39 帕斯卡 1磅力/平方英寸=6 894.76 帕斯卡 1巴=100.0 千帕斯卡 1标准大气压=101.325 千帕斯卡
Moment/Torque 力矩/扭矩	$1\,\text{lbf in}=0.112\,985\,\text{Nm}$ $1\,\text{lbf ft}=1.355\,82\,\text{Nm}$	1磅力英寸=0.112 985 牛顿米 1磅力英尺=1.355 82 牛顿米

（续表）

量的名称	换算系数（英）	换算系数（中）
Energy/Heat/Work 能量/热量/功	1 hp·h=2.684 52 MJ 1 therm=105.506 MJ 1 Btu=1.055 06 kJ 1 kWh=3.6 MJ	1[英制]马力小时=2.684 52 兆焦耳 1撒姆=105.506 兆焦耳 1英热单位=1.055 06 千焦耳 1千瓦小时=3.6 兆焦耳
Heat Flow Rate 热流率	1 Btu/h=0.293 071 W	1英热单位/小时=0.293 071 瓦
Power 功率	1 hp(550 ft lbf/s)=745.700 W	1 hp[英制]马力=745.700 瓦
Kinematic Viscosity 动态黏度	1 ft^2/s=929.03 stokes=0.092 903 m^2/s	1平方英尺/秒=929.03 斯托克斯=0.092 903 平方米秒
Specific Enthalpy 比焓	1 Btu/ft^3=37.258 9 kJ/m^2	1英热单位/立方英尺=37.258 9 千焦耳/平方米
Plane Angle 平面角	1 rad=57.295 8degrees 1 degree=0.017 453 3 rad=1.1111 grade 1 second=4.848 14×10^{-6} rad=0.000 3 grade 1 minute=2.908 88×10^{-4} rad=0.018 5 grade	1弧度=57.295 8度 1度=0.017 453 3弧度=1.1111梯度 1秒=4.848 14×10^{-6}弧度=0.000 3梯度 1分=2.908 88×10^{-4}弧度=0.018 5梯度
Velocity of Rotation 转速	1 r/min=0.104 720 rad/s	1转/分钟=0.104 720 弧度/秒